Pesticides in Surface and Ground Water of the San Joaquin–Tulare Basins, California: Analysis of Available Data, 1966 Through 1992

By JOSEPH L. DOMAGALSKI

Prepared in cooperation with the
National Water-Quality Assessment Program

U.S. GEOLOGICAL SURVEY WATER-SUPPLY PAPER 2468

U.S. DEPARTMENT OF THE INTERIOR
BRUCE BABBITT, Secretary

U.S. GEOLOGICAL SURVEY
Gordon P. Eaton, Director

1997

For sale by the
U.S. Geological Survey
Information Services
Box 25286
Federal Center
Denver, CO 80225

Library of Congress Cataloging in Publication Data

Domagalski, Joseph L.
Pesticides in surface and ground water of the San Joaquin-Tulare Basins, California: analysis of available data, 1966 through 1992 / by Joseph L. Domagalski ; prepared in cooperation with the National Water-Quality Assessment Program.
p. cm. -- (U.S. Geological Survey water-supply paper ; 2468) Includes bibliographical references.
1. Pesticides--Environmental aspects--California--San Joaquin Valley. 2. Pesticides--Environmental aspects--California--Tulare Lake Watershed. 3. Water--Pollution--California--San Joaquin Valley. 4. Water-Pollution--California--Tulare Lake Watershed. I. National Water Quality Assessment Program (U.S.) II. Title. III. Series: Geological Survey water-supply paper ; 2468.
TD427.P35D66 1997
363.73'84--dc20 96-5832
CIP
ISBN-0-607-86814-7

FOREWORD

The mission of the U.S. Geological Survey (USGS) is to assess the quantity and quality of the earth resources of the Nation and to provide information that will assist resource managers and policymakers at Federal, State, and local levels in making sound decisions. Assessment of water-quality conditions and trends is an important part of this overall mission.

One of the greatest challenges faced by water-resources scientists is acquiring reliable information that will guide the use and protection of the Nation's water resources. That challenge is being addressed by Federal, State, interstate, and local water-resource agencies and by many academic institutions. These organizations are collecting water-quality data for a host of purposes that include: compliance with permits and water-supply standards; development of remediation plans for specific contamination problems; operational decisions on industrial, wastewater, or water-supply facilities; and research on factors that affect water quality. An additional need for water-quality information is to provide a basis on which regional- and national-level policy decisions can be based. Wise decisions must be based on sound information. As a society we need to know whether certain types of water-quality problems are isolated or ubiquitous, whether there are significant differences in conditions among regions, whether the conditions are changing over time, and why these conditions change from place to place and over time. The information can be used to help determine the efficacy of existing water-quality policies and to help analysts determine the need for and likely consequences of new policies.

To address these needs, the U.S. Congress appropriated funds in 1986 for the USGS to begin a pilot program in seven project areas to develop and refine the National Water-Quality Assessment (NAWQA) Program. In 1991, the USGS began full implementation of the program. The NAWQA Program builds upon an existing base of water-quality studies of the USGS, as well as those of other Federal, State, and local agencies. The objectives of the NAWQA Program are to:

- Describe current water-quality conditions for a large part of the Nation's freshwater streams, rivers, and aquifers.
- Describe how water quality is changing over time.
- Improve understanding of the primary natural and human factors that affect water-quality conditions.

This information will help support the development and evaluation of management, regulatory, and monitoring decisions by other Federal, State, and local agencies to protect, use, and enhance water resources.

The goals of the NAWQA Program are being achieved through ongoing and proposed investigations of 60 of the Nation's most important river basins and aquifer systems, which are referred to as study units. These study units are distributed throughout the Nation and cover a diversity of hydrogeologic settings. More than two-thirds of the Nation's freshwater use occurs within the 60 study units and more than two-thirds of the people served by public water-supply systems live within their boundaries.

National synthesis of data analysis, based on aggregation of comparable information obtained from the study units, is a major component of the program. This effort focuses on selected water-quality topics using nationally consistent information. Comparative studies will explain differences and similarities in observed water-quality conditions among study areas and will identify changes and trends and their causes. The first topics addressed by the national synthesis are pesticides, nutrients, volatile organic compounds, and aquatic biology. Discussions on these and other water-quality topics will be published in periodic summaries of the quality of the Nation's ground and surface water as the information becomes available.

This report is an element of the comprehensive body of information developed as part of the NAWQA Program. The program depends heavily on the advice, cooperation, and information from many Federal, State, interstate, Tribal, and local agencies and the public. The assistance and suggestions of all are greatly appreciated.

Robert M. Hirsch

Robert M. Hirsch
Chief Hydrologist

CONTENTS

FIGURES

TABLES

CONVERSION FACTORS, VERTICAL DATUM, AND ADDITIONAL ABBREVIATIONS

Multiply	By	To obtain
acre	4,047	square meter
acre-foot (acre-ft)	1,233	cubic meter
acre-foot per year (acre-ft/yr)	1,233	cubic meter per year
foot (ft)	0.3048	meter
gallon (gal)	3.785	liter
gallons per day (gal/d)	0.003785	cubic meters per day
inch (in.)	25.4	millimeter
inch per hour (in/h)	25.4	millimeter per hour
mile (mi)	1.609	kilometer
pound (lb)	0.4536	kilogram
pounds per acre (lb/acre)	112.1	kilograms per square kilometer
square mile (mi^2)	2.590	square kilometer

Temperature is given in degrees Fahrenheit (°F) or degrees Celsius (°C) and can be converted by using the following equations:

$$°F = 1.8\ (°C) + 32$$
$$°C = 5/9\ (°F) - 32$$

VERTICAL DATUM

***Sea level*:** In this report, "sea level" refers to the National Geodetic Vertical Datum of 1929—a geodetic datum derived from a general adjustment of the first-order level nets of both the United States and Canada, formerly called Sea Level Datum of 1929.

ADDITIONAL ABBREVIATIONS

g/L grams per liter
µg/L micrograms per liter
µg/kg micrograms per kilogram

Pesticides in Surface and Ground Water of the San Joaquin–Tulare Basins, California: Analysis of Available Data, 1966 Through 1992

By Joseph L. Domagalski

Abstract

Available pesticide data (1966-92) for surface and ground water were analyzed for the San Joaquin–Tulare Basins, California, one of 60 large hydrologic systems being studied as part of the National Water-Quality Assessment Program of the U.S. Geological Survey. Most of the pesticide data were for the San Joaquin Valley, one of the most intensively farmed and irrigated areas of the United States. Data were obtained from the Storage and Retrieval data base of the U.S. Environmental Protection Agency, the water-quality data base of the U.S. Geological Survey, and from data files of State agencies.

Pesticides detected in surface water include organochlorine pesticides, organophosphate pesticides, carbamate pesticides, and triazine herbicides. Pesticides detected in ground water include triazine and other organonitrogen herbicides and soil fumigants. Surface-water data indicate seasonal patterns for the detection of organophosphate and carbamate pesticides, which are attributed to their use on almond orchards and alfalfa fields. Organochlorine pesticides were detected primarily in river-bed sediments. Concentrations detected in bed sediments of the San Joaquin River near Vernalis are among the highest of any major river system in the United States. Patterns and timing of pesticide use indicate that pesticides might be present in surface-water systems during most months of a year.

The most commonly detected pesticide in ground water is the soil fumigant, dibromochloropropane. Dibromochloropropane, used primarily on vineyards and orchards, was detected in ground water near the city of Fresno. Triazine and other organonitrogen herbicides were detected near vineyards and orchards in the same general locations as the detections of dibromochloropropane. Pesticides were detected in ground water of the east side of the valley floor, where the soils are sandy or coarse-grained, and water-soluble pesticides with long environmental half-lives were used. In contrast, fewer pesticides were detected in ground water of the west side of the valley, where soils generally are finer grained.

INTRODUCTION

The quality of the Nation's ground- and surface-water resources is being affected by numerous human and natural processes. Available data generally are inadequate to assess the status and trends in water quality of large regions of the Nation. In 1991, after a pilot phase, the U.S. Geological Survey (USGS) began to implement the National Water-Quality Assessment Program (NAWQA) that integrates information about water quality at a wide range of spatial scales, from local to national, and focuses on water-quality conditions that affect large areas of the Nation or occur frequently within numerous small areas. The NAWQA Program is described in detail in a report by Hirsch and others (1988).

In 1991, the San Joaquin–Tulare Basins study unit in California was selected as one of the first 20 NAWQA study units. Each study unit is a defined hydrologic basin or group of basins that has been selected for studies of water-quality issues. These issues include concentrations of pesticides, nutrients, and trace elements found in surface and ground water. During the first phase of NAWQA, evaluation of pesticide and nutrient concentrations is the highest priority. An important first step for each study is to review what is already known about each of these issues. In particular, the study design and selection of sampling locations for each NAWQA study unit will be highly influenced by the availability and interpretation of available information on the priority constituents. Retrospective analysis includes review and analysis of available water-quality data on a particular constituent or set of constituents. Retrospective reports on nutrients and pesticides are planned for each study unit as one of the first major products for the NAWQA Program.

Agriculture is a major land use within the San Joaquin Valley and is of critical importance to the Nation's supply of certain commodities. Some commodities, such as almonds, are grown only in California. The San Joaquin Valley also supplies a significant portion of the Nation's grapes, raisins, oranges, and other fruits. Pesticides have been used extensively in the San Joaquin Valley and pose a potential threat to both ground and surface water.

Data were obtained from the Storage and Retrieval (STORET) data base of the U.S. Environmental Protection Agency, the water-quality data base of the USGS, and from data files of various California State agencies. The earliest records evaluated for this report were for 1966; therefore, the period of record is 1966 to 1992.

The purpose of this report is to present a review and an analysis of available data on pesticides in surface and ground water for the San Joaquin–Tulare Basins study unit, California. Pesticide data for surface water are insufficient to calculate loads or yearly trends, but can be used to assess geographic and seasonal occurrence of select pesticides and to relate their occurrence to their agricultural use in the study unit. Pesticide data for ground water were sufficient to assess the presence of pesticides in ground water adequately and to relate the presence of pesticides in ground water to pesticide use in the basins.

DESCRIPTION OF STUDY AREA

The boundary of the study unit (fig. 1) is defined by the drainage divides of the streams that enter the San Joaquin and Tulare Basins from the Sierra Nevada and Coast Ranges. This boundary coincides with the boundaries of basin numbers 18040000 and 18030000 as shown on the hydrologic unit map of California (U.S. Geological Survey, 1978).

Physiographic and Geologic Setting

The San Joaquin–Tulare Basins study unit has a drainage area of 28,500 mi^2 in three major physiographic provinces of central California: the Sierra Nevada, the San Joaquin Valley, and the Coast Ranges (fig. 1). The study unit can be further separated into the San Joaquin Basin to the north and the hydrologically closed Tulare Basin to the south. The San Joaquin Valley is a flat structural basin bounded by the Sierra Nevada to the east, the Coast Ranges to the west, the Tehachapi Mountains to the south, and the Sacramento-San Joaquin Delta to the north. Land-surface altitudes of the valley rise from several feet above sea level in the north to 1,000 ft at the top of the dissected older alluvium in the southeast. The Sierra Nevada attains a maximum altitude of 14,495 ft at Mount Whitney, the highest point in the conterminous United States.

The bedrock geology of the Sierra Nevada contrasts sharply with that of the Coast Ranges. The Sierra Nevada is composed primarily of pre-Tertiary granitic rocks and is separated from the valley by a foothill belt of Mesozoic and Paleozoic marine rocks and Mesozoic metavolcanic rocks along the northern one-third of the boundary (California Division of Mines and Geology, 1959a, 1959b, 1965a, 1965b, 1966, 1967, and 1969). The Coast Ranges have a core of Franciscan complex of Late Jurassic to Late Cretaceous or Paleocene age and ultramafic rocks of Meso-

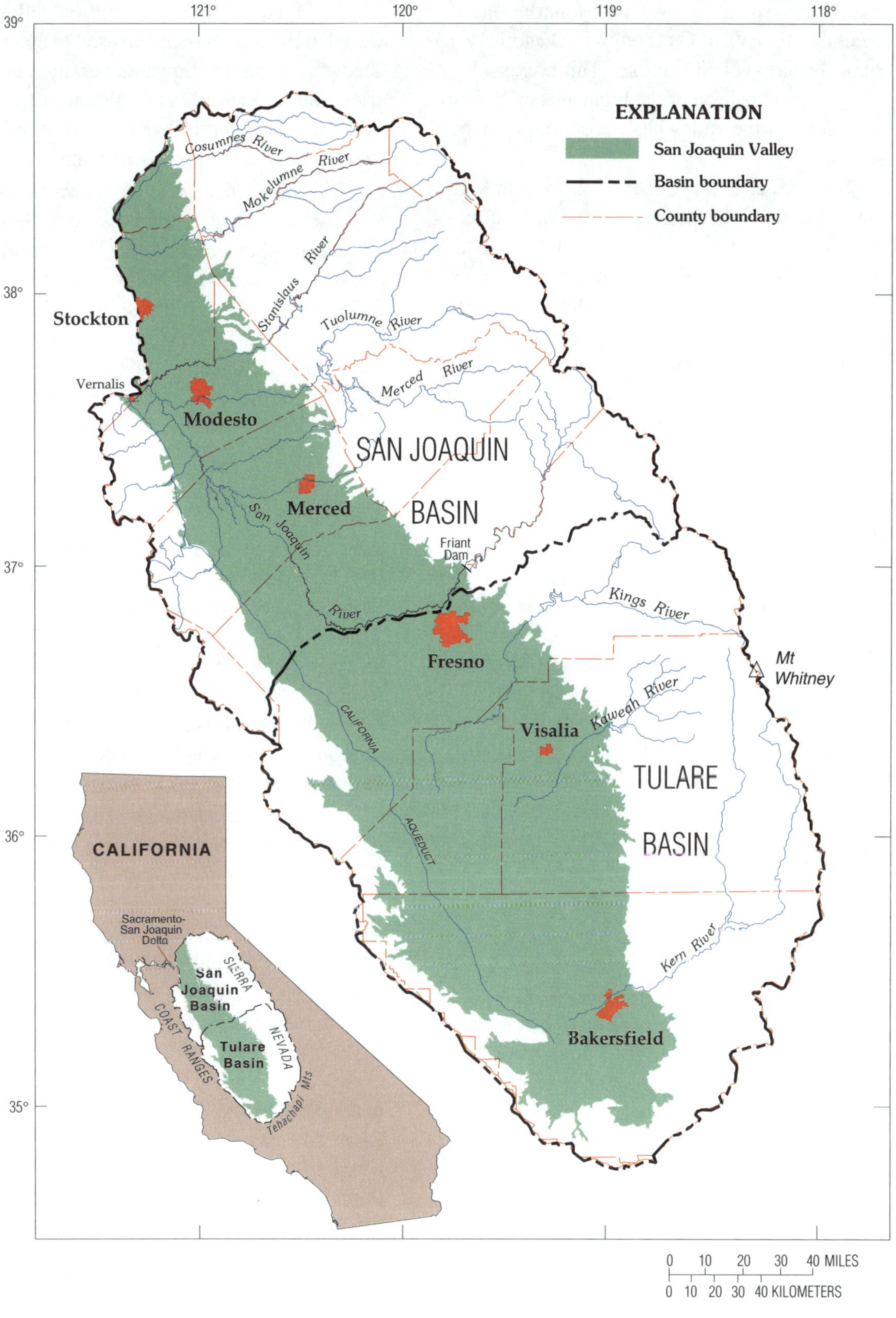

Figure 1. Location of San Joaquin–Tulare Basins study unit.

zoic age. These rocks are overlain by marine and continental sediments of Cretaceous to Quaternary age and some Tertiary volcanic rocks. This contrast between the composition of the highlands on the east and west sides of the valley has a profound influence on water quality.

The surficial geology of the San Joaquin Valley is shown in figure 2. Sediments of the San Joaquin Valley consist of interlayered gravel, sand, silt, and clay derived from the adjacent mountains and deposited in alluvial-fan, floodplain, flood-basin, lacustrine, and marsh environments. Lithology and texture of the sediments reflect their source area and manner of deposition. Sediments derived from the Coast Ranges are finer grained than those derived from the Sierra Nevada.

Climate

The valley floor is arid to semiarid, with a mean annual rainfall ranging from 5 in. in the south to 15 in. in the north. Almost all rainfall is in the winter. Annual precipitation in the Sierra Nevada, most of which is snow in some areas, averages up to 80 in. Temperatures in the valley are extremely variable with mean daily temperatures ranging from about 35 to 102°F.

Surface-Water Hydrology

The surface-water hydrology of the San Joaquin–Tulare Basins study unit has been radically modified by development of water resources. The timing of surface-water flow from the Sierra Nevada into the valley has been greatly affected by the construction of dams. Maximum river flows were in late winter or early spring, and minimum river flows were in late summer and early autumn. Now, every major river that enters the valley from the Sierra Nevada has one or more reservoirs.

The total mean annual flow into the San Joaquin Valley from the Sierra Nevada drainages, mostly runoff from the Sierra Nevada snowpack, is 8,840,000 acre-ft/yr (Nady and Larragueta, 1983). The total reservoir storage capacity of the dams on the Stanislaus, the Tuolumne, the Merced, the San Joaquin, the Kings, and the Kern Rivers (fig. 1) is about 7.5 million acre-ft (10 acre-ft equals about 3.3 million gallons). Water stored in the reservoirs is released to meet downstream needs for irrigation, power generation, in-stream fisheries, and recreational uses. When sufficient storage is available, the reservoirs retain part of the snowmelt runoff peaks to reduce downstream flooding.

In sharp contrast to the Sierra Nevada tributaries, most streams that drain the Coast Ranges are intermittent or ephemeral and contribute an insignificant amount of water to the valley. The total mean annual flow from the Coast Ranges, including the Tehachapi Mountains (fig. 1), is estimated to be 92,600 acre-ft/yr (Nady and Larragueta, 1983), which represents about 1 percent of the total surface water that enters the San Joaquin Valley from the adjacent highlands.

The San Joaquin River is the only surface-water outlet from the study unit. The water quality of the San Joaquin River is of critical interest because the river flows into the Sacramento-San Joaquin Delta (fig. 1). The delta is the source of irrigation water to farms in the San Joaquin Valley and the California Aqueduct (fig. 1), which in turn supplies part of the drinking water for 15 million people in southern California. The delta also is a key ecological resource for several important fish, such as striped bass and salmon, and for waterfowl species that are present at various stages of their life cycles.

Friant Dam on the San Joaquin River (fig. 1), at the boundary between the Sierra Nevada foothills and the valley floor, retains most of the Sierra Nevada runoff to the upper San Joaquin River. As a result, flow in most of the first 132 mi of river below Friant Dam is intermittent. Accordingly, flow in the lower reaches depends on inflow from the major east-side tributaries (the Merced, the Tuolumne, and the Stanislaus Rivers), agricultural drainage, and return flows. For example, during one low-flow period, only about 17 percent of the total flow measured on the San Joaquin River downstream from the Stanislaus River originated upstream from the Merced River; 65 percent of the flow was derived from the three major east-side tributaries (Clifton and Gilliom, 1989). The remaining 18 percent of the flow was derived predominantly from ground-water inflow, agricultural drainage, and return flows. The same study found that during high-flow conditions, about 44 percent of the flow of the San Joaquin River below its confluence with the

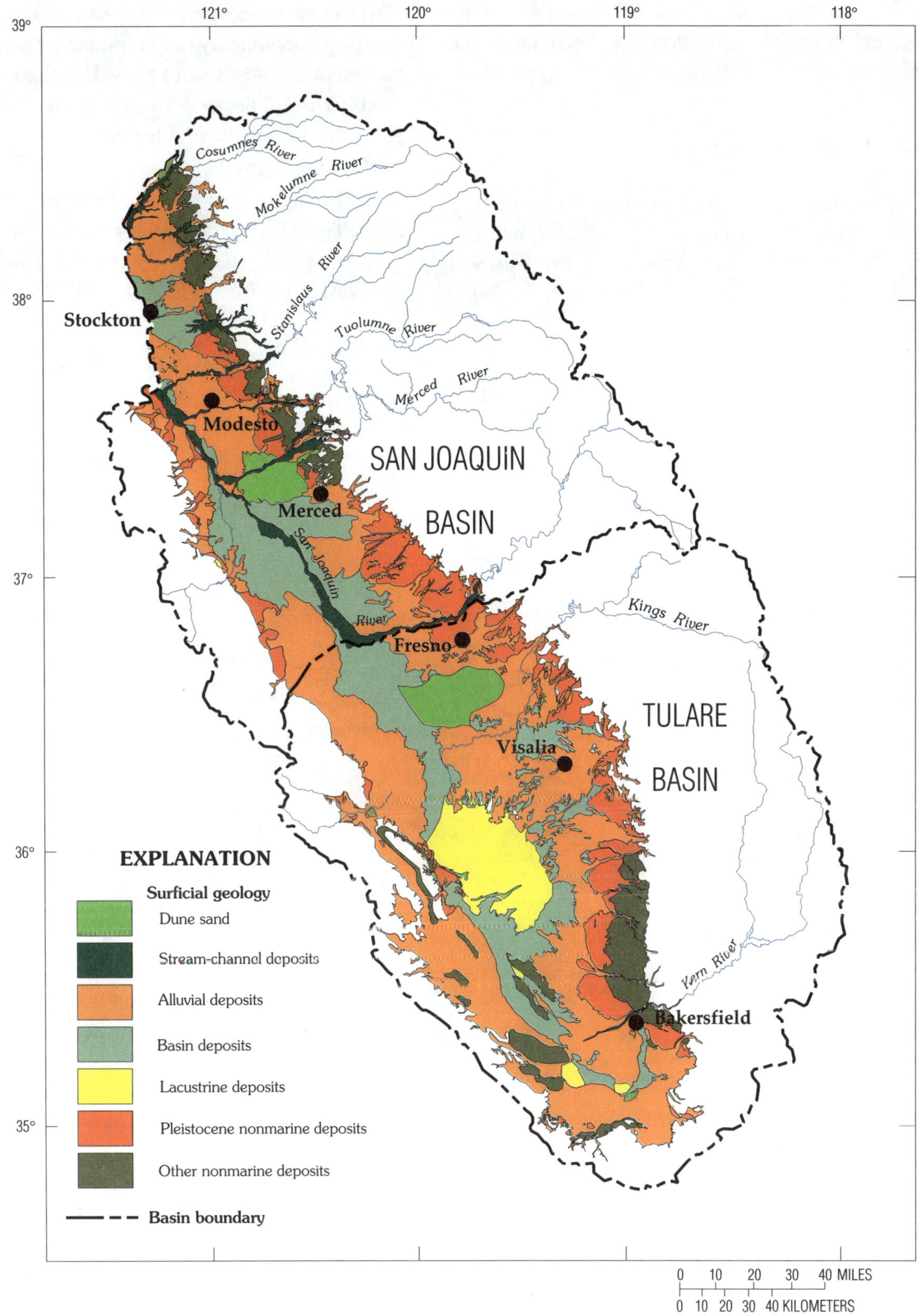

Figure 2. Generalized surficial geology of the San Joaquin Valley, California (modified from California Division of Mines and Geology, 1959a, 1959b, 1965a, 1965b, 1966, 1967, and 1969).

Stanislaus River originated upstream from the Merced River, and 53 percent of the flow was derived from the three major east-side tributaries.

Ground-Water Hydrology

Ground water is withdrawn from an aquifer in the San Joaquin Valley that averages about 1,400 ft thick. The aquifer is composed of numerous lenses of gravel, sand, silt, and clay. Sediments in the west-side alluvial fans are coarsest at the heads of the fans and consist predominantly of fine-grained silts and clays in the rest of the west side of the valley (Laudon and Belitz, 1991). Sediments derived from the Sierra Nevada generally are coarser grained than those derived from the Coast Ranges.

The most areally extensive fine-grained unit in the valley is the Corcoran Clay Member of the Tulare Formation of Pleistocene age (fig. 3). This unit underlies most of the west side of the valley at a depth of as

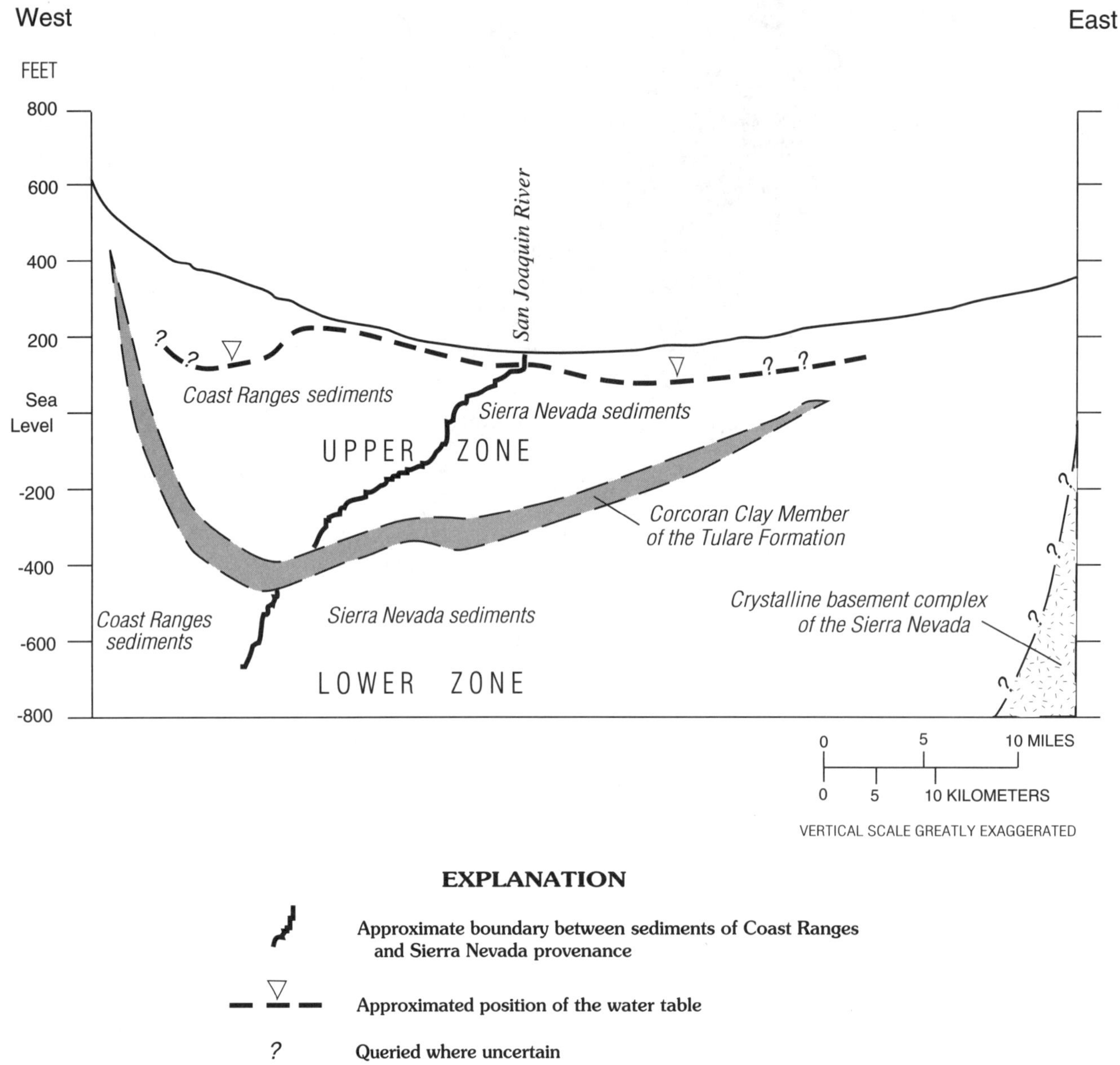

Figure 3. Generalized geohydrologic section through the central part of the San Joaquin Valley, California (from Davis and others, 1959).

much as 650 ft and attains a thickness of at least 160 ft in the center of the Tulare Basin (Davis and others, 1959; Page, 1986). Prior to ground-water resources development, part of the aquifer that underlies the Corcoran Clay Member was confined; however, recent hydraulic connection of the zones above and below the clay, attributed to the large number of wells that are perforated above and below the clay, has changed the ground-water-flow paths and reduced the effectiveness of the Corcoran Clay Member as a confining unit. In parts of the valley, the Corcoran Clay Member separates zones with distinctly different water chemistries (Davis and Poland, 1957; Davis and others, 1959; Davis and Coplen, 1989).

Because of the differences in hydraulic head and water chemistry above and below the Corcoran Clay, the clay generally was believed to separate the aquifer system into unconfined or semiconfined zones (above the clay) and a confined zone (below the clay). Williamson and others (1989) presented a different view; they found that hydraulic gradients across the clay were similar to those at shallower depth, indicating that the Corcoran Clay was no more significant on a regional scale than were other less easily identifiable fine-grained lenses. Contrasts in water chemistry above and below the Corcoran Clay make it an important marker in any study of ground-water quality, regardless of its influence in the physical flow system.

Population and Land Use

The population of the study unit was 2.7 million in 1990 (U.S. Bureau of Census, 1990) and is growing rapidly as small, formerly agricultural towns in the northeast part of the study unit become "bedroom communities" for the San Francisco metropolitan area. About 46 percent of the residents are in the four largest urban areas—Fresno, Stockton, Bakersfield, and Modesto (fig. 1); most of the rest are in small farming communities in the San Joaquin Valley. The Sierra Nevada and Coast Ranges adjacent to the valley are sparsely populated.

The relative proportions of land uses in the study unit, based on mid-1970's data (U.S. Geological Survey, 1986), are 40 percent forest, 25 percent cropland and pasture, 23 percent rangeland, 6 percent orchards, 3 percent barren land, 2 percent urban area, and less than 1 percent wetland. Almost the entire San Joaquin Valley (about 10 million acres) is agricultural land, whereas most of the Sierra Nevada is forested. About 70 percent of the agricultural land is irrigated. The history of agriculture in the valley extends back to the 1870's. Large-scale farming and the amount of farmland expanded after World War I and continued to increase as water-distribution systems improved. Additional areas were irrigated with water imported from northern California.

The combination of abundant water and the long growing season results in an exceptionally productive agricultural economy in the San Joaquin Valley. In 1987, California produced 10.2 percent of the total value of agricultural production in the United States, 49 percent of which was generated in the San Joaquin Valley. Major products include livestock and livestock products (35 percent), fruits and nuts (33 percent), cotton (13 percent), vegetables (6.5 percent), and hay and grains (6 percent). Other agricultural uses account for 6.5 percent (San Joaquin Valley Drainage Program, 1990). The distribution of crops generally reflects the distribution of soil texture and chemistry. About 43 percent of the irrigated land on the west side of the valley is planted in cotton, which is salt tolerant. In contrast, 31 percent of the cropland on the east side of the valley is planted in fruit and nut trees, which are intolerant of salinity and some trace elements, and accounts for 92 percent of the total acreage of these crops in the valley.

Water Use

Approximately 95 percent of water use in the San Joaquin Valley is for irrigated agriculture. In 1990, the total consumptive use was 12.1 million acre-ft; about 58 percent was from surface-water, and 42 percent was from ground water (W.E. Templin, U.S. Geological Survey, written commun., 1992).

The pattern of surface-water use varies greatly in different parts of the valley. The abundant Sierra Nevada runoff on the east side of the valley, regulated by the system of reservoirs described previously, is widely used for agriculture and drinking-water. Most surface water used for irrigation on the west side of the valley is imported from northern California using aqueducts constructed by the Bureau of Reclamation and the State of California. These projects provide irrigation water for about 1 million acres.

In the 1950's and 1960's, the large rate of ground-water pumping led to extensive overdrafts and declining water levels, causing land subsidence in some areas on the west side of the valley. The availability of imported water in the western San Joaquin Basin and parts of the Tulare Basin resulted in a decrease in ground-water pumping (Belitz and Heimes, 1990). This shift in water use contributed to a rise in ground-water levels that resulted in a water table within 5 ft of land surface in 800,000 acres in the valley. Most of this area requires subsurface drainage, which lowers the water table sufficiently to prevent buildup of salts in the soils and subsequent crop damage. Much of the ground-water system is still overdrafted because of continued extensive pumping. For example, in 1990, there was a net loss of 3.1 million acre-ft of ground water from storage in the San Joaquin Valley; in 1991 (the fifth consecutive year of below-normal rainfall), the net loss was projected to be about 8 million acre-ft (Carl Hauge, California Department of Water Resources, written commun., 1992).

ENVIRONMENTAL FRAMEWORK FOR WATER-QUALITY ASSESSMENT

The environmental framework refers to natural and man-made features of a hydrologic system, such as geology and land use, that provide a unifying framework for comparative assessments of water quality within and among NAWQA study units. Water-quality data for the NAWQA Program are interpreted or discussed according to subdivisions of the study unit on the basis of hydrogeologic-physiographic features and land use. The San Joaquin–Tulare Basins study unit has three major physiographic provinces: the Sierra Nevada, the Coast Ranges, and the San Joaquin Valley (fig. 4). Alluvial fan and basin deposits of the San Joaquin Valley also are shown in figure 4. Most, if not all, pesticide data reviewed in this report are restricted to the San Joaquin Valley because most of the population and agriculture, and therefore water use and activities that affect water quality, are in this area. Moreover, most of the surface-water data focuses on the part of the valley in the San Joaquin Basin, specifically the perennial reach of the San Joaquin River, because (1) the perennial San Joaquin River is the only surface-water outlet of the basin, (2) the water quality of the San Joaquin River affects water quality in the Sacramento-San Joaquin Delta, and (3) the Tulare Basin does not have an integrated surface-water-flow system.

Physiography and Land Use

As described previously, the San Joaquin Valley can be characterized as a sedimentary basin with flat topography that rises from sea level where the San Joaquin River enters the Sacramento-San Joaquin Delta to about 1,000 ft in altitude at the southern end. The valley can be divided into three subareas on the basis of geomorphology, sediment source, and depositional characteristics: alluvial fans east of the depositional axis, basin, and alluvial fans west of the depositional axis. Although the depositional axis of the valley has shifted during geologic time, the east-side alluvial fans are dominated by sediments derived from the Sierra Nevada, and the west-side alluvial fans are dominated by sediments derived from the Coast Ranges. Sediments in the basin are a mixture of Sierra Nevada and Coast Ranges sources reworked and deposited in stream channels, shallow lakes, and as overbank deposits in flood basins.

Physiographic provinces of the study unit are shown in figure 4. Environmental settings, areas of relatively uniform physiography, and land use are shown in figure 5. Basin deposits constitute an important setting for consideration of ground-water quality, but will not be considered a separate setting for design of the surface-water assessment because they generally are not drained by perennial streams; hence, there is no appropriate site to monitor the effect of runoff from the basin areas on the perennial stretch of the San Joaquin River. Where extensive basin areas drain to the river, much of the flow that reaches the river originates as agricultural runoff and tailwater from adjacent upgradient alluvial fans and, thus, would not reflect the water quality of runoff from the basin areas.

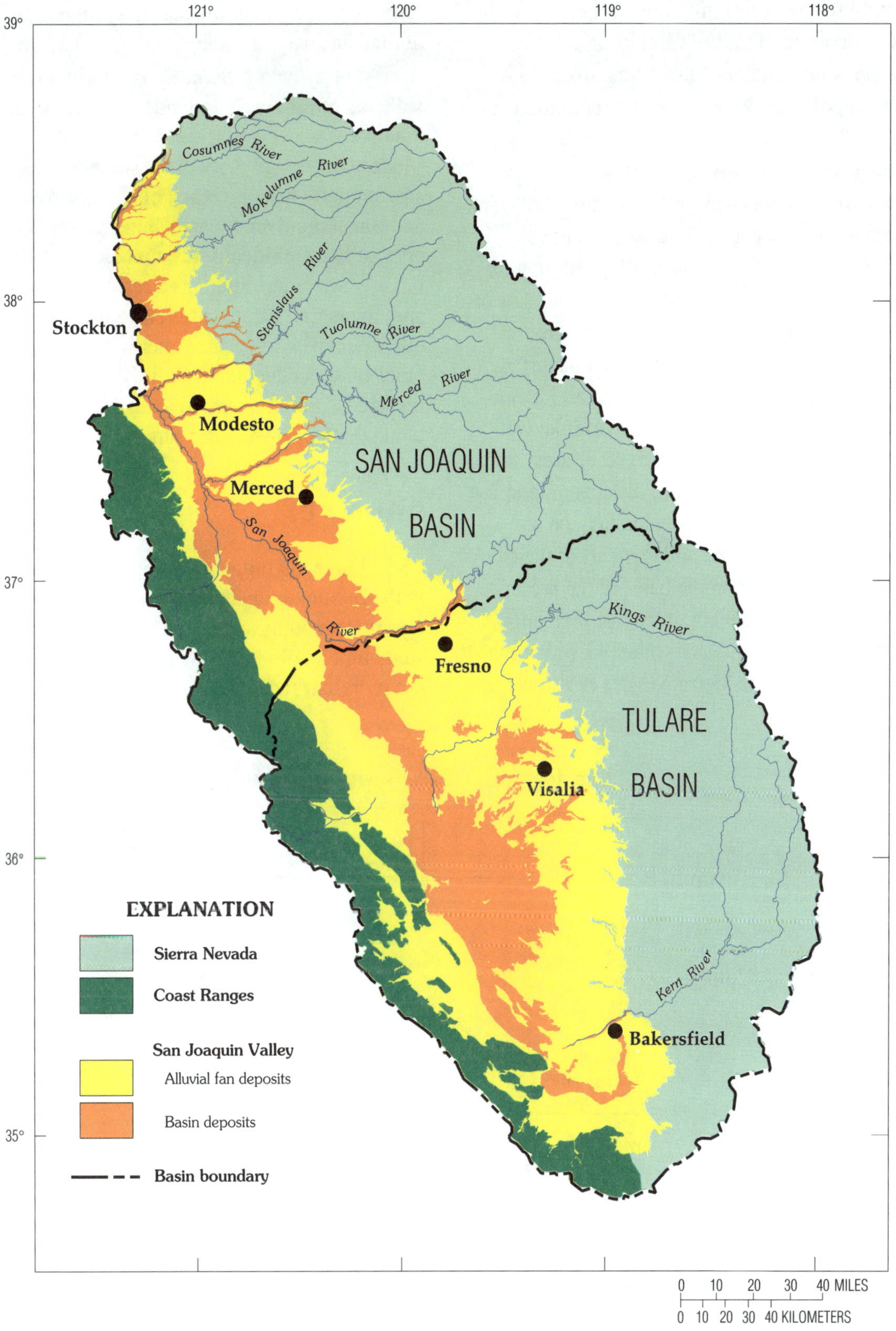

Figure 4. Physiographic provinces in the San Joaquin–Tulare Basins study unit and delineation of alluvial and basin deposits in the San Joaquin Valley, California.

Land-use categories in figure 5 reflect the dominance of agriculture in the area and its anticipated influence on water quality. At this time, focus is on assessing the contrasts in water-quality conditions in the major categories of agricultural land use: row crops, field crops, orchards, and dairies.

Row crops include various vegetables and some fruits, such as strawberries. Field crops generally include nonedible crops such as alfalfa, cotton, field corn, and sugar beets. In spite of the large number of dairies, this land use is considered only for ground-water studies. Dairies are not considered in surface-water studies because runoff is regulated and, therefore, limited. About 95 percent of the original wetlands in the valley were converted to other uses, so the area of wetlands is not significant. Effects of land use on water quality in the Sierra Nevada may be an important issue in the future, and this part of the study unit could be subdivided on the basis of the presence or absence of logging, mining, intensive recreation, and reservoirs. At present, assessment of water quality in subdivisions of the Sierra Nevada is a lower priority than understanding the effect of various land uses in the San Joaquin Valley.

The effect of urban areas on water quality in the valley is of relatively low interest because agriculture dominates water-quality and water-resource issues. Modesto is the only city in the study unit located along a perennially flowing reach of a major river. Fresno and Bakersfield are located along reaches of major rivers that had significant flows under natural conditions but are now intermittent as a result of flow regulation and diversion. Effects of urban areas on ground-water quality may be of greater interest because all major cities in the valley currently rely on ground-water resources, and thus, urban areas potentially could contaminate their own water supply.

Pesticide Use

Large quantities of pesticides are used each year in the San Joaquin–Tulare Basins study unit. The reported amounts applied in the eight counties of the study unit in 1988 are shown in table 1. These data do not represent the actual total of pesticides applied

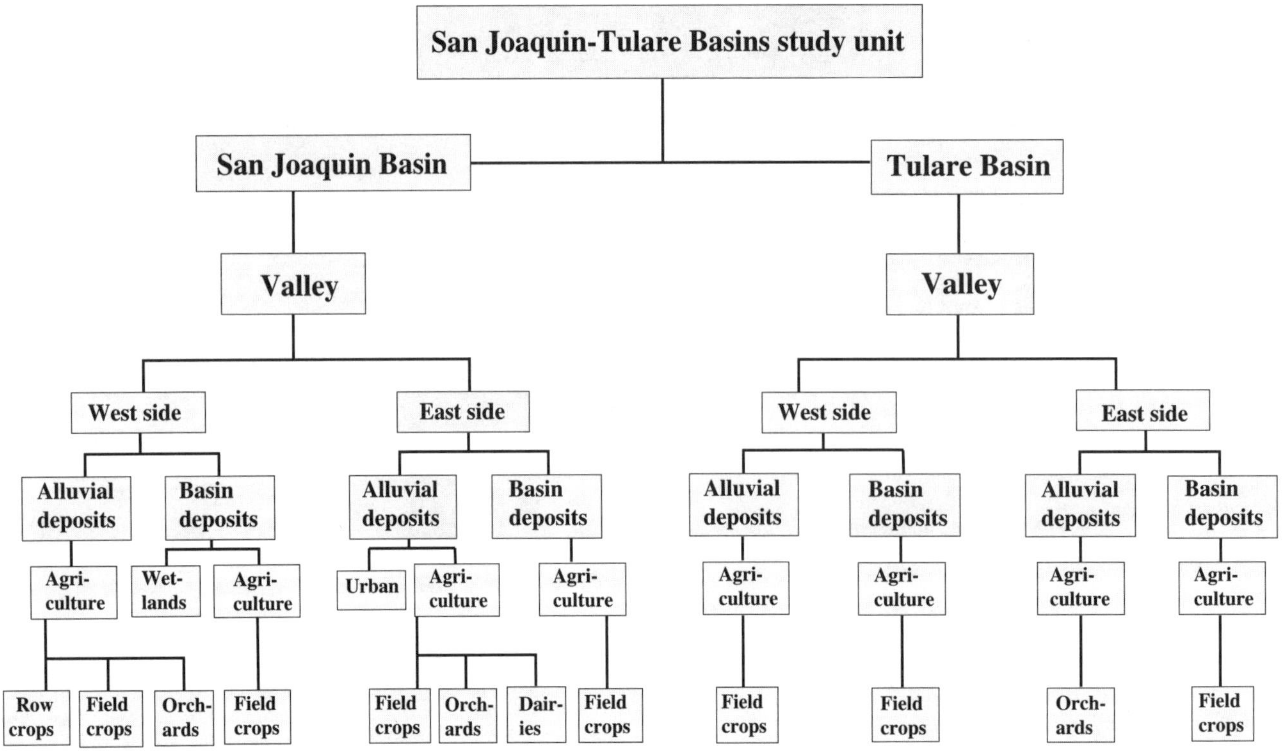

Figure 5. Environmental settings and land use in the San Joaquin–Tulare Basins study unit, San Joaquin Valley, California.

Table 1. Total pesticide applications in the San Joaquin–Tulare Basins study unit, San Joaquin Valley, California, in 1988, by county

[lb, pound. Data from California Department of Pesticide Regulation, 1988]

County	Total pesticide application (lb)
Fresno	16,375,690
Kern	10,997,150
San Joaquin	6,096,779
Tulare	5,610,710
Merced	5,006,569
Kings	3,712,147
Stanislaus	3,582,684
Madera	3,514,259
Total	54,895,988

in the study unit. The State of California reporting requirements in 1988 included use of all restricted pesticides and all pesticide applications by licensed applicators (California Department of Pesticide Regulation, 1988). Use of nonrestricted pesticides by private farming operations were not reported. Data shown in table 1 and in subsequent tables do not include the applications of herbicides for weed control along roadways or other nonagricultural uses. A rule change in 1990 now requires that all pesticide applications be reported.

Approximately 350 pesticides were used in the study unit during 1988. The 60 most heavily applied pesticides account for 96 percent of the total pesticide use in the study unit (table 2). Despite their relatively

Table 2. Pesticides applied in the San Joaquin–Tulare Basins study unit, San Joaquin Valley, California, in 1988

[lb, pound. Data from California Department of Pesticide Regulation, 1988]

Pesticide	Total pesticide application (lb)	Pesticide	Total pesticide application (lb)
Sulfur	16,723,660	Endosulfan	183,724
1,3-dichloropropene	8,960,924	Copper sulfate, basic	179,340
Sodium chlorate	3,814,121	Naled	179,160
Methyl bromide	3,804,462	Pendimethalin	165,802
Petroleum oil, unclassified	3,762,001	Diuron	160,309
Propargite	1,385,593	Captan	157,975
Ethyl parathion	874,101	Malathion	157,244
Chlorpyrifos	789,247	Mancozeb	154,543
S,S,S-tributylphosphorotrithioate (DEF)	761,371	Sodium cacodylate	154,037
Cryolite	749,953	Petroleum hydrocarbons	152,623
Profenofos	688,741	Formentanate hydrochloride	129,681
Methomyl	612,176	Maneb	125,621
Copper hydroxide	612,077	Chlorine	125,559
Ethephon	543,466	Phorate	124,170
Chloropicrin	506,107	2,4-D dimethylamine salt	98,149
Glyphosphate	449,187	Petroleum distillates, aromatic	96,800
Carbaryl	449,102	Phosmet	95,646
Paraquat dichloride	444,658	Merphos	95,288
Trifluralin	357,734	Oryzalin	89,441
Mineral oil	352,559	Carbofuran	87,826
Azinphos-methyl	340,476	Metam-sodium	85,020
Dicofol	322,766	Oxyfluorfen	80,372
Aldicarb	310,034	Endothall (mono-alkylamine salt)	80,355
Ziram	302,198	Xylene	79,130
Acephate	297,465	Simazine	77,372
Dimethoate	292,866	Copper	74,924
Methamidophos	280,313	Chlorthall-dimethyl	69,818
Methidathion	240,086	MCPA dimethylamine salt	65,663
Cyanazine	204,372	Monocrotophos	65,183
Diazinon	185,169	Cypermethrin	63,663

low use, some of the remaining 4 percent, 290 pesticides, have the potential to affect water quality adversely.

The types of pesticides used reflect the types of crops grown in the study unit. After cotton and sugar beets, the most common crops are fruits and nuts. Major pesticides include fungicides, such as sulfur, and soil fumigants, such as 1,3-dichloropropene and methyl bromide. Some of the most heavily applied pesticides are used exclusively on cotton. Examples include sodium chlorate and S,S,S-tributylphosphorotrithioate (DEF), used as cotton defoliants prior to the autumn harvest, and profenofos, an organophosphate pesticide. Soil fumigants, used on various crops, accounted for 23 percent of total pesticides used in the study unit during 1988. Soil fumigants eliminate soil nematodes, but must be used prior to planting because of their phytotoxicity. The amount of soil fumigants used per acre typically is higher than most other types of pesticides. The amount of methyl bromide applied per acre is highly variable, with a mean application of 332 lb/acre and a median of 56 lb/acre for all crop uses. In contrast, the mean application of trifluralin, a pre-emergent herbicide, is 4.5 lb/acre for all crop uses.

Table 3. Agricultural uses of pesticides in the San Joaquin–Tulare Basins study unit, San Joaquin Valley, California, in 1988

[lb, pound. Data from California Department of Pesticide Regulation, 1988]

Crop	Total pesticide application (lb)
Cotton	11,269,100
Grapes	8,314,009
Sugar beets	7,697,213
Almonds	5,708,581
Tomatoes	5,027,580
Open land	1,808,454
Oranges	1,709,462
Carrots	1,402,323
Alfalfa	1,192,644
Strawberries	1,146,500
Peaches	1,038,770
Fallow farm land	699,219
Sweet potatoes	689,702
Roses	537,957
Plums	517,937
Walnuts	490,408
Pistachios	457,107
Nectarines	419,592
Beans	415,305
Potatoes	367,708
Ornamentals	355,252
Corn	329,985
Head lettuce	293,729
Dried fruit	267,919
Onions	267,247
Olives	220,122
Wheat	202,646

The large variety of pesticides used reflects the diverse agricultural practices of the San Joaquin Valley, as well as the need to use pesticides for various purposes throughout the year. Major agricultural uses of these pesticides in the study unit during 1988 are shown in table 3.

Pesticides were applied to 129 different crops in 1988. The 27 agricultural uses listed in table 3 account for 96 percent of the total pesticide use for that year. The remaining 4 percent of pesticides were used on 102 crops; pesticide use on some of those may adversely affect water quality.

Pesticides are used in the San Joaquin Valley throughout much of the year (table 4). Many herbicides are applied from November to March, and pesticides are applied to almond and other stone-fruit orchards when trees are dormant, from December to February. Pesticides may enter surface-water systems with runoff from rain-saturated fields because those sprays usually are applied during the rainy season. Insecticides are sprayed on alfalfa fields during March. Although March storms generally are less intense than January and February storms, March rainfall can result in the mobilization of alfalfa pesticides to surface-water systems. The amounts of herbicides and insecticides used on cotton are high; therefore, the potential for transport to surface-water systems is high.

During the irrigation season, April through September, insecticide use is high, but herbicide use is relatively low. Rainfall usually is not sufficient to cause runoff, and irrigated fields are drained. Drainwater discharges into surface-water systems, which eventually enter the San Joaquin River.

Table 4. Seasonal pesticide use in the San Joaquin Valley, California, in 1988

[lb, pound]

Application period or crop type	Compound	Total pesticide application (lb)	Months of use
Winter herbicides	Trifluralin	174,035	November to March.
	Paraquat dichloride	160,160	Do.
	Diuron	147,567	Do.
	Oryzalin	81,906	Do.
	2,4-D	80,000	Do.
	Oxyfluorfen	76,499	Do.
	Glyphosphate	75,591	Do.
	Simazine	68,291	Do.
	MCPA, dimethylamine salt	56,705	Do.
	Chlorthall-dimethyl	54,840	Do.
	Metolachlor	1,104	Do.
	Atrazine	95	Do.
Dormant orchard spray	Petroleum oil	2,629,847	December to February.
	Mineral oil	272,697	Do.
	Diazinon	96,772	Do.
	Methidathion	41,988	Do.
	Chlorpyrifos	10,414	Do.
Alfalfa insecticides	Chlorpyrifos	55,474	March to April.
	Phosmet	55,240	Do.
	Diazinon	46,676	Do.
	Carbofuran	43,272	Do.
	Malathion	38,923	Do.
	Methomyl	2,179	Do.
	Propargite	196	Do.
Spring/summer herbicides	Glyphosphate	97,635	April to September.
	Trifluralin	80,685	Do.
	Paraquat dichloride	36,697	Do.
	Metolachlor	10,301	Do.
	MCPA, dimethylamine salt	8,943	Do.
	Alachlor	8,407	Do.
	Atrazine	7,187	Do.
	Chlorthall-dimethyl	6,475	Do.
	Diuron	3,479	Do.
	Simazine	2,005	Do.
Spring/summer insecticides	Cryolite	741,237	Do.
	Methomyl	504,554	Do.
	Carbaryl	427,565	Do.
	Chlorpyrifos	356,705	Do.
	Azinphos-methyl	335,444	Do.
	Diazinon	262,904	Do.
	Dimethoate	256,511	Do.
	Naled	174,823	Do.
	Endosulfan	168,300	Do.
	Methidathion	165,181	Do.
	Formentanate hydrochloride	126,895	Do.

Table 4. Seasonal pesticide use in the San Joaquin Valley, California, in 1988—Continued

Application period or crop type	Compound	Total pesticide application (lb)	Months of use
Spring/summer insecticides	Petroleum distillate aromatic	90,272	April to September.
	Phorate	71,555	Do.
	Malathion	63,120	Do.
	Phosmet	41,158	Do.
	Carbofuran	11,382	Do.
Cotton herbicides (includes pre-harvest defoliants)	S,S,S-tributylphosphorotrithioate (DEF)	761,044	September to October.
	Pendimethalin	333,641	November to March.
	Glyphosphate	234,167	March.
	Paraquat dichloride	242,909	September to October.
	Cyanazine	178,683	November to February and June to July.
	Trifluralin	103,020	November to March.
	Merphos	95,127	September to October.
	Endothall	75,573	Do.
Cotton insecticides/nematicides	Profenofos	687,473	May to September.
	Propargite	529,874	Do.
	Chlorpyrifos	330,116	Do.
	Aldicarb	281,400	April to June.
	Dicofol	252,988	May to August.
	Methamidophos	211,767	June to September.
	Acephate	192,624	January to August.
	Cypermethrin	52,259	July to September.
	Dimethoate	28,954	May to August.
	Methomyl	26,487	May to September.

SOURCES OF PESTICIDE DATA

Sources of pesticide data for surface- and ground-water sites are listed in table 5. Three databases were used for the retrospective analysis of surface-water data: the STORET data base of the U.S. Environmental Protection Agency, the water-quality data base of the USGS, and a data base from the California Regional Water Quality Control Board. Ground-water data were available from the well-inventory data base of the California Department of Pesticide Regulation (Maes and others, 1991) and the water-quality data base of the USGS.

Surface Water

Analyses of samples from tile drain discharges comprise most of the surface-water data from the

Table 5. Sources of pesticide data for surface- and ground-water sites in the San Joaquin–Tulare Basins study unit, San Joaquin Valley, California

Data base or agency	Number of sites	Period of record
Surface-water sites		
STORET[1]	113	1966–87
U.S. Geological Survey	31	1972–90
California Regional Water Quality Control Board	20	1991–92
Ground-water sites		
California Department of Pesticide Regulation[2]	4,623	1971–90
U.S. Geological Survey	362	1984–90

[1]U.S. Environmental Protection Agency Storage and Retrieval water-quality data base.

[2]California Department of Pesticide Regulation, well-inventory data base (Maes and others, 1991).

STORET data base. Data also are available for surface-water sites sampled for pesticides in the San Joaquin–Tulare Basins study unit (fig. 6; table 6).

Although the STORET data base contains data from a number of State and Federal agencies, most surface-water-quality data from the study area were contributed to the STORET data base by the California Department of Water Resources. For the Tulare Basin, a significant amount of water-quality data in the STORET data base was collected by the U.S. Army Corps of Engineers. Data of the USGS are included in STORET; however, they were analyzed separately from the data of other agencies. Quality assurance and quality-control practices of the USGS are well documented, whereas those of other agencies could not be fully documented. Because of the long time frame and participation of several agencies, evaluation of all quality assurance programs, including those of the analytical laboratories, was not feasible for this study.

Data from the USGS include contributions from the San Joaquin Valley Drainage Program, the National Stream Quality Accounting Network (NASQAN), and the USGS Toxic Substances Hydrology Program. The Toxic Substances Hydrology Program and NASQAN contributed data for only one site, the San Joaquin River near Vernalis. The San Joaquin Valley Drainage Program and NASQAN samples were obtained by compositing water from equal-discharge- or equal-width-increment sampling of the channel. Teflon or stainless-steel equipment was used, and samples were analyzed by the U.S. Geological Survey laboratory in Arvada, Colorado, according to methods published in a report by Wershaw and others (1987). Bed sediment samples were collected in depositional areas of fine-grained sediment. Bed material was sampled with a BMH-53 core sampler or BMH-60 bed sampler (Guy and Norman, 1970), or a K.B.-type core sampler (Greeson and others, 1977). Samples from the Toxic Substances Hydrology Program were collected near the center of flow of the channel and were analyzed for selected pesticides by the USGS laboratory in Sacramento, California (Crepeau and others, 1991).

The data base of the California Regional Water Quality Control Board was created from a study done in cooperation with the USGS and includes data from an individual investigation of pesticide concentrations and biotoxicity of water at a variety of sites. Data from the California Regional Water Quality Control Board were interpreted independently of data from the STORET and USGS databases. The California Regional Water Quality Control Board used a three-species biotoxicity test that requires 3 days (Foe and Connor, 1991) to analyze grab samples collected near the stream banks. After 3 days, samples that showed significant toxicity relative to a control, as defined by California Regional Water Quality Control Board tests, were shipped to the USGS laboratory in Arvada, Colorado, where they were stored at 4°C until they could be analyzed for pesticides.

Ground Water

The purposes of the well-inventory data base of the California Department of Pesticide Regulation are to "centralize reliable information on the occurrence of nonpoint-source contamination of ground water by the agricultural use of pesticides and to facilitate graphical, numerical, and spatial analyses of the data" (Maes and others, 1991). The data base is maintained principally to facilitate the regulation of the use of pesticides in the State of California; therefore, data are rigorously screened by the California Department of Pesticide Regulation. Quality assurance of this data is considered adequate for analytical chemistry and location of wells. The sampling techniques used for individual wells are unknown.

Most of the ground-water-quality data from the study unit were contributed to the California Department of Pesticide Regulation by the California Department of Health Services. Other agencies contributing to the California Department of Pesticide Regulation data base include the California Department of Pesticide Regulation's Environmental Hazards and Assessment Program, the California Department of Water Resources, and some county agencies. Systematic regional or subregional samplings were not done for any particular pesticide. The California Department of Health Services measures well samples for chemical constituents and monitors ground-water quality because ground water is a major source of drinking water for communities in the study unit.

Table 6. Surface-water sites sampled for pesticides by various agencies in the San Joaquin–Tulare Basins study unit, San Joaquin Valley, California

[Station no.: Unique number for each site. Fifteen-digit numbers are based on the geographic location of the site, beginning with the latitude and longitude; eight-digit numbers refer to frequently sampled sites along a major stream; the number is assigned in downstream order. See figures 6, 20, 21, and 31 for site locations]

Site no.	Station no.	Location name	Site no.	Station no.	Location name
		San Joaquin Basin			
1	364400120202600	Fresno Slough at State Highway 180.	18	370822120433500	San Joaquin River at Turner Island at Erreca Road.
2	364639120241601	Tile Drain FBH2016.	19	371317120464600	Mariposa Bypass-Slough.
3	364718120221800	San Joaquin River below Mendota Pool at Bass Avenue.	20	371354120541000	Santa Fe Canal at Gun Club Road.
4	364822120271701	Tile Drain FBH4045.	21	11261100	Salt Slough at State Highway 165 near Stevinson.
5	364915120283301	Tile Drain FBH5056.	22	11260800	Bear Creek near Stevinson.
6	11255760	Firebaugh Wasteway at Helm Canal Road.	23	11262900	Mud Slough near Gustine.
7	365128120301901	Tile Drain BVS7402.	24	11260815	San Joaquin River near Stevinson.
8	365128120391401	Tile Drain HMH7516.	25	371636120575200	Los Banos Creek at State Highway 140.
9	11255800	San Joaquin River at Firebaugh Bridge.	26	11261500	San Joaquin River at Fremont Ford Bridge.
10	365337120445301	Tile Drain DPS1367.	27	371903120585400	Newman Wasteway near State Highway 33.
11	365440120412501	Tile Drain DPS2535.	28	11274000	San Joaquin River near Newman.
12	365524120411801	Panoche Drain.	29	11272500	Merced River near Stevinson.
13	365526120441301	Tile Drain DPS3465.	30	372406120573000	Turlock Irrigation District drain lateral 6.
14	365630120391401	Tile Drain DPS4616.	31	372236121032400	Orestimba Creek at State Highway 33.
15	365630120451802	Camp 13 Slough at Head B0096300.	32	11274554	Spanish Grant Drain.
16	11256000	San Joaquin River near Dos Palos.	33	11274560	Turlock Irrigation District drain lateral 5.
17	370635120350400	San Joaquin River at Washington Road Bridge.			

Table 6. Surface-water sites sampled for pesticides by various agencies in the San Joaquin–Tulare Basins study unit, San Joaquin Valley, California—Continued

Site no.	Station no.	Location name
San Joaquin Basin—Continued		
34	372805121042301	Tile Drain CLG6520.
35	372903121053301	Tile Drain CLG7551.
36	372907121053701	Tile Drain CLG7651.
37	11274570	San Joaquin River at Patterson Bridge near Patterson.
38	373156121035600	Turlock Irrigation District drain lateral 3.
39	373048121093000	Del Puerto Creek at State Highway 33.
40	373334121090200	San Joaquin River at Laird Park.
41	373548121134701	Tile Drain WLY6652.
42	373648121123001	Tile Drain WLY7845.
43	11290000	Tuolumne River at Modesto.
44	373747121125200	Ingram Creek at River Road.
45	373842121131800	Hospital Creek at River Road.
46	373701121121100	Ingram/Hospital Creek.
47	11290500	San Joaquin River at Maze Road near Modesto.
48	11303000	Stanislaus River at Ripon.
49	374209121103800	Stanislaus River at Caswell State Park.
50	11303500	San Joaquin River near Vernalis.
51	374043121183801	Tile Drain VNS3733.
52	374119121185201	Tile Drain VNS4535.
Tulare Basin		
1	350702119021201	Tile Drain COC8221.
2	351231119180401	Tile Drain MOK5829.
3	353304119302901	Tile Drain STC3505.
4	353527119364401	Tile Drain STC6467.
5	354512118252401	Kern River at Kernville.
6	360306119352401	Tile Drain CCN3550.
7	360456119384301	Tile Drain ERR5712.
8	360632119400901	Tile Drain ERR7525.
9	360813119433201	Tile Drain GSY0855.
10	362448119004201	Kaweah River above Lake Kaweah.
11	364948119200601	Kings River above Pine Flat Lake.

Ground-water data were collected by the USGS as part of the San Joaquin Valley Drainage Program or the Regional Aquifer Systems Analysis program. The San Joaquin Valley Drainage Program made primarily reconnaissance-level measurements of shallow ground water in the central part of the western San Joaquin Valley. The wells sampled were water-table observation wells that mostly were less than 20 ft deep. In addition, a regional-scale study of the aquifers that supply water to the San Joaquin Valley was completed by the Regional Aquifer Systems Analysis program (Domagalski and Dubrovsky, 1991, 1992). The wells sampled were domestic or irrigation wells.

The wells were purged prior to sampling, and at least three casing volumes were pumped. Water samples were taken after pH, temperature, and specific conductance had stabilized and then were analyzed for volatile organic compounds, triazine herbicides, organophosphate pesticides, carbamate pesticides, and chlorophenoxy acid herbicides. Details of the analytical method are given in a report by Wershaw and others (1987).

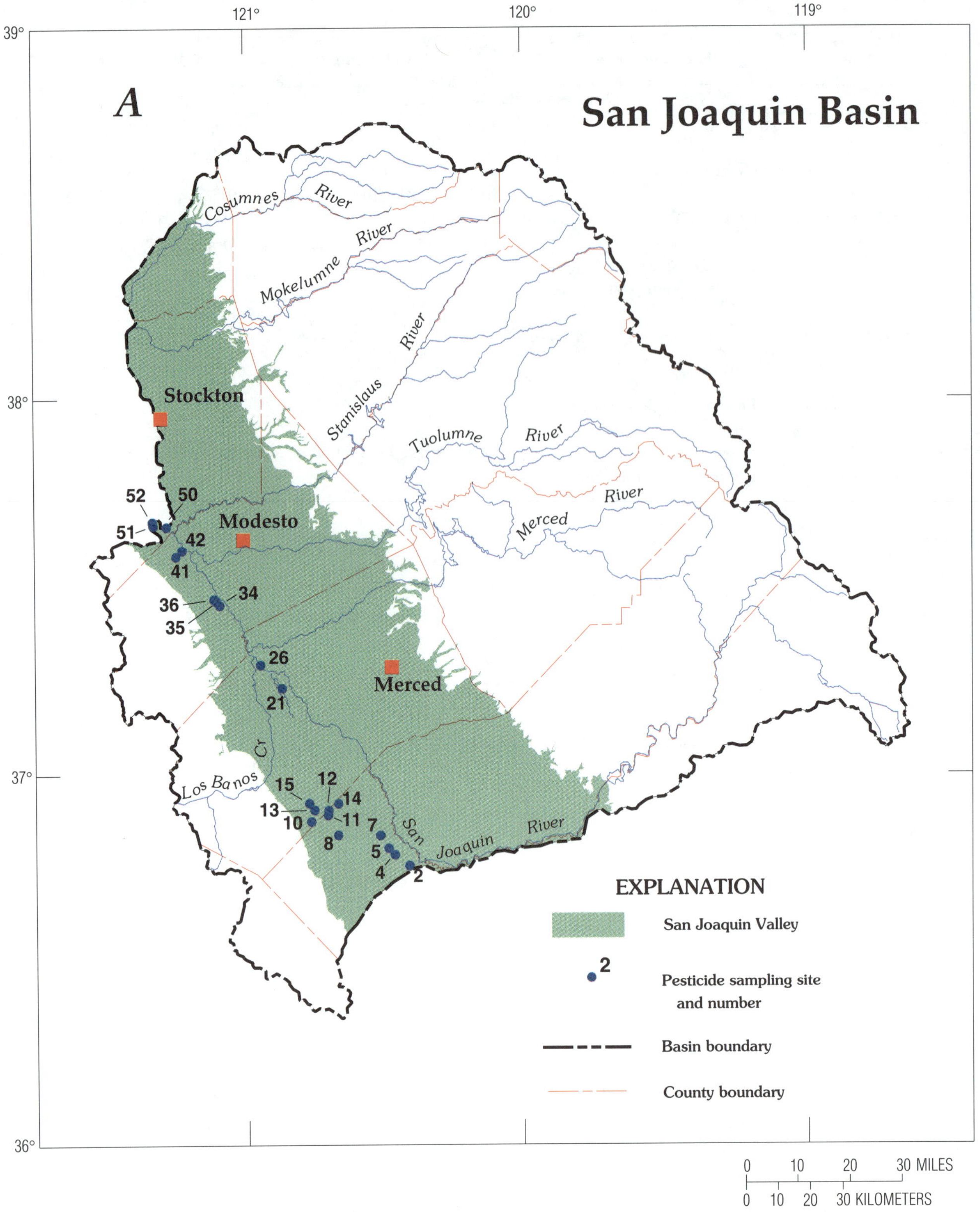

Figure 6. Locations of selected sites sampled for pesticides in the San Joaquin–Tulare Basins study unit, San Joaquin Valley, California. *A*, San Joaquin Basin; *B*, Tulare Basin. (Compiled from U.S. Environmental Protection Agency's STORET data base. See table 6 for station numbers and descriptions.)

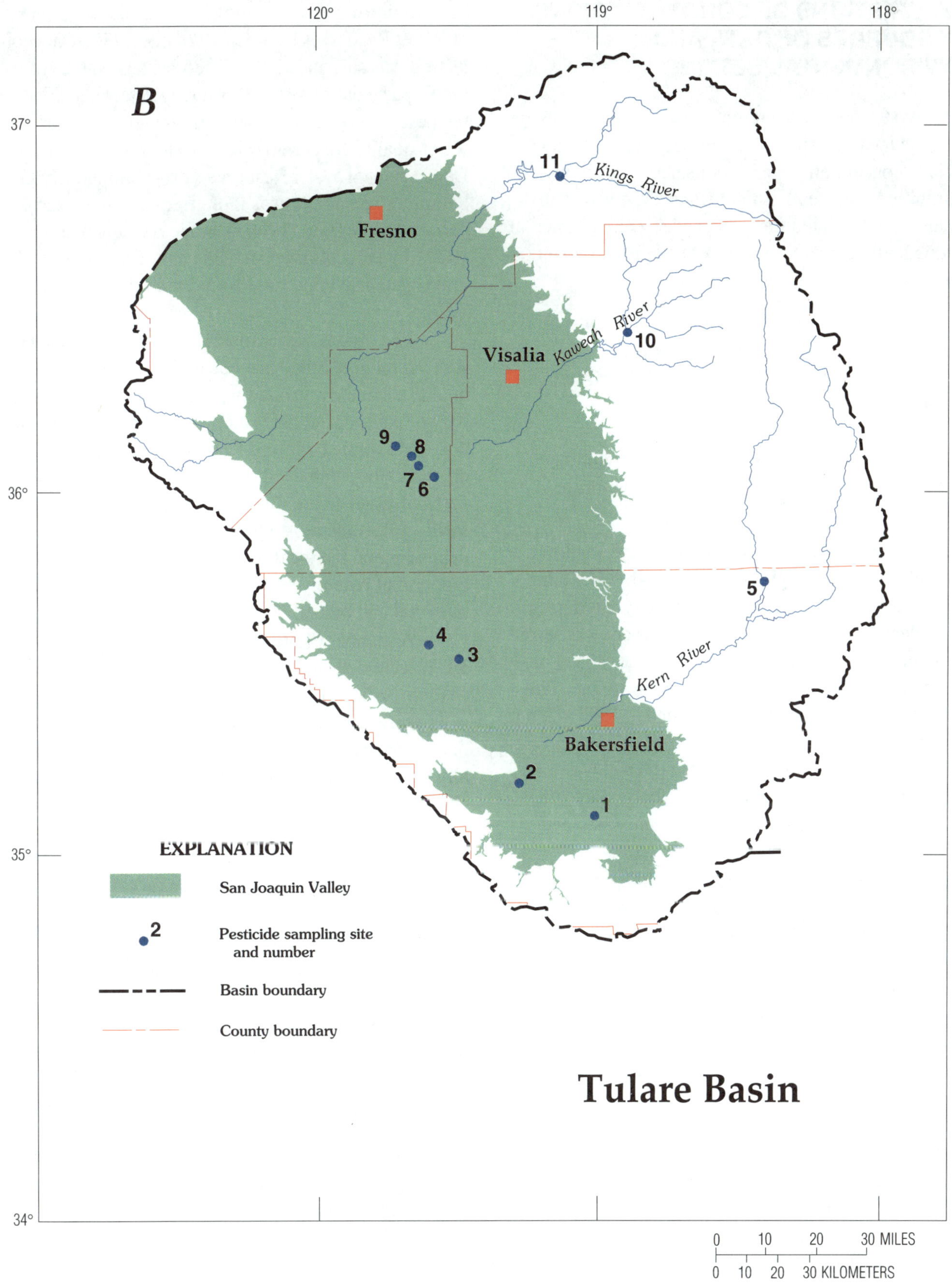

Figure 6—Continued.

DESCRIPTIONS OF CONCENTRATIONS BY SOURCES OF DATA AND ENVIRONMENTAL SETTING

Water-quality data from individual databases are presented and interpreted in this section. The data include concentrations of various pesticides in water, bottom sediment, and ground water. The water-quality data are presented for each available environmental setting and include both surface- and ground-water data.

Surface Water—U.S. Environmental Protection Agency, Storage and Retrieval Data Base

San Joaquin Basin: Valley, West Side, Alluvial Deposits Environmental Setting—Agriculture (Field Crops)

Tile drains were installed to remove shallow ground water from regions where a high water table intersects the root zone of crops, thereby interfering with plant growth. Initially, tile drainwater was stored in basin areas for seasonal waterfowl habitat and discharged to the San Joaquin River in the spring. This tile drainwater storage was suspended in 1985 when it was determined that the tile drainwater generally contained high concentrations of selenium and was toxic to waterfowl (Gilliom and others, 1989). Since then, tile drainwater is discharged into streams that feed directly into the San Joaquin River.

Selected organochlorine pesticide concentrations in water samples collected from tile drains are shown in figure 7. DDT was detected in most of the tile-drain samples. The median DDT concentrations ranged from 0.009 µg/L at tile drain BVS7402 (fig. 7*C*; site 7, fig. 6*A*) to 0.18 µg/L at the Panoche Drain (fig. 7*G*; site 12, fig. 6*A*). Other detected organochlorine pesticides include dieldrin, lindane, and toxaphene. The median concentrations of toxaphene were highest (1.5 and 1.0 µg/L) at the Panoche Drain (fig. 7*G*; site 12, fig. 6*A*) and tile drain WLY7845 (fig. 7*H*; site 42, fig. 6*A*). The relatively higher concentrations may be due partly to the fact that toxaphene is not an individual compound but rather a complex mixture of chlorinated camphenes.

Most of the pesticide data from tile-drain samples were collected by the San Joaquin Valley Drainage Monitoring Program of the California Department of Water Resources between 1966 and 1985. Data collection was suspended after 1985 because the California Department of Water Resources determined that the pesticide concentrations were insignificant for water-quality concerns (Michael McGinnis, California Department of Water Resources, oral commun., 1993). Sampling apparently was more frequent in the early part of the program (1960's) and less frequent after 1974. Organochlorine pesticide concentrations in tile-drain systems varied little from the mid-1960's to 1974.

Tile drainwater is subsurface water and, under some conditions, may provide an indication of the quality of shallow ground water. However, the chemistry of a tile-drainwater sample may be different from that of a sample collected from an observation well drilled to the same depth because of preferential flow paths of water to tile drains. Tile drains do not necessarily collect the most recently recharged water, but depending on the local hydrology, may collect a combination of recently recharged and older ground water (Deverel and others, 1989).

A number of organochlorine pesticides, such as DDT, were detected in tile drainwater by the San Joaquin Valley Drainage Monitoring Program. DDT is unlikely to be detected in ground water because it has a tendency to sorb to soil particles and is not usually mobile in the subsurface. Detections of organochlorine pesticides may be attributed to various factors. Land use in the regions of the tile drains is largely field crop farming. Cotton is one of the major commodities grown. Pesticide use on cotton is high, primarily to control the cotton weevil, and cotton farming currently represents one of the major uses of insecticides in the San Joaquin–Tulare Basins study unit. Prior to its ban in 1971, DDT was used heavily in this region of the San Joaquin Valley; however, actual records of pesticide use for that period do not exist. DDT and other organochlorine pesticide residues probably persist in soils because of the long half-lives of those pesticides. The detection of DDT in the subsurface possibly is due to mobilization attributable to partitioning among the solid, water, and dissolved organic-carbon phases. Solubility of DDT and other organochlorine compounds can be enhanced by partitioning to the dissolved organic carbon of ground water (Chiou and others, 1987; Johnson-Logan and others, 1992). Subsurface mobility is enhanced if

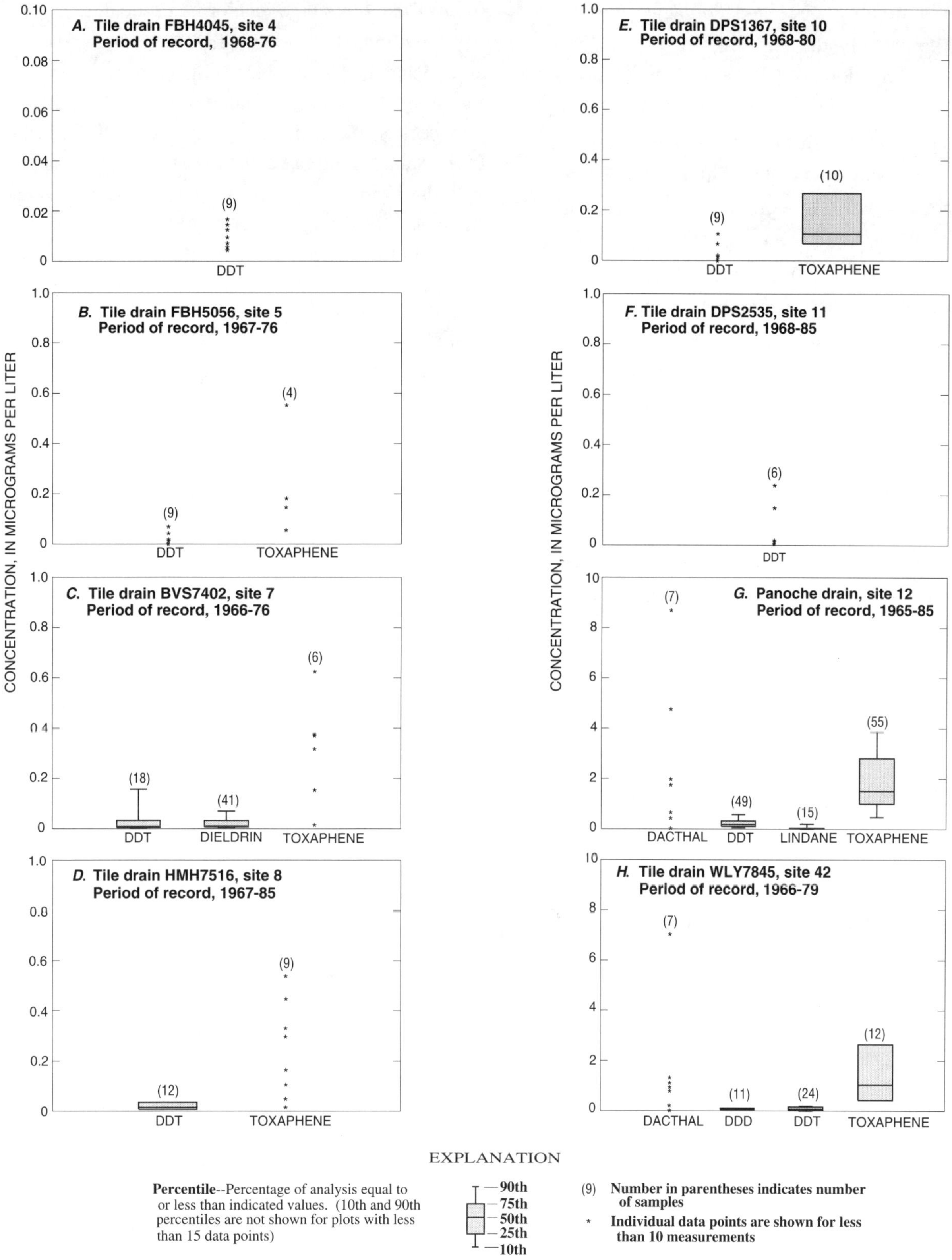

Figure 7. Concentrations of selected organochlorine pesticides in water samples collected from tile drains, San Joaquin Basin, California: Valley, west side, alluvial deposits environmental setting—agriculture (field crops). (Compiled from U.S. Environmental Protection Agency's STORET data base. See figure 6*A* for site locations.)

DDT or other organochlorine pesticides migrate with the natural dissolved or colloidal organic carbon.

Alternatively, the detections of DDT might be an artifact of the drain construction. Tile-drain installation requires excavation of the soil. When the soil is backfilled at the site, surface soil is mixed with relatively cleaner soil from depth. Therefore, the drain is in contact or in very close proximity to soil containing DDT or other pesticides, and the detections might be due to slow desorption of the pesticides from these soils or erosion of these soils into the tile drains. A complete hydrological interpretation would require a detailed study of each individual drain source. However, the detections of pesticides in this area were consistent with the land use and the use of large amounts of DDT and other organochlorine pesticides.

Individual data points for DDT concentrations at the Panoche Drain (site 12, fig. 6*A*) are shown in figure 8. The Panoche Drain was selected for this display because of the available data. The DDT concentrations may be affected by seasonality, particularly with the 1966 and 1967 data where the sampling was frequent. DDT concentrations were highest in the summer months when the contribution from tile drains is highest. The Panoche Drain actually receives a combination of water from both tile drain and surface runoff, although the contribution of tile drainwater probably is higher than for surface water. However, the surface-water contribution probably is highest in

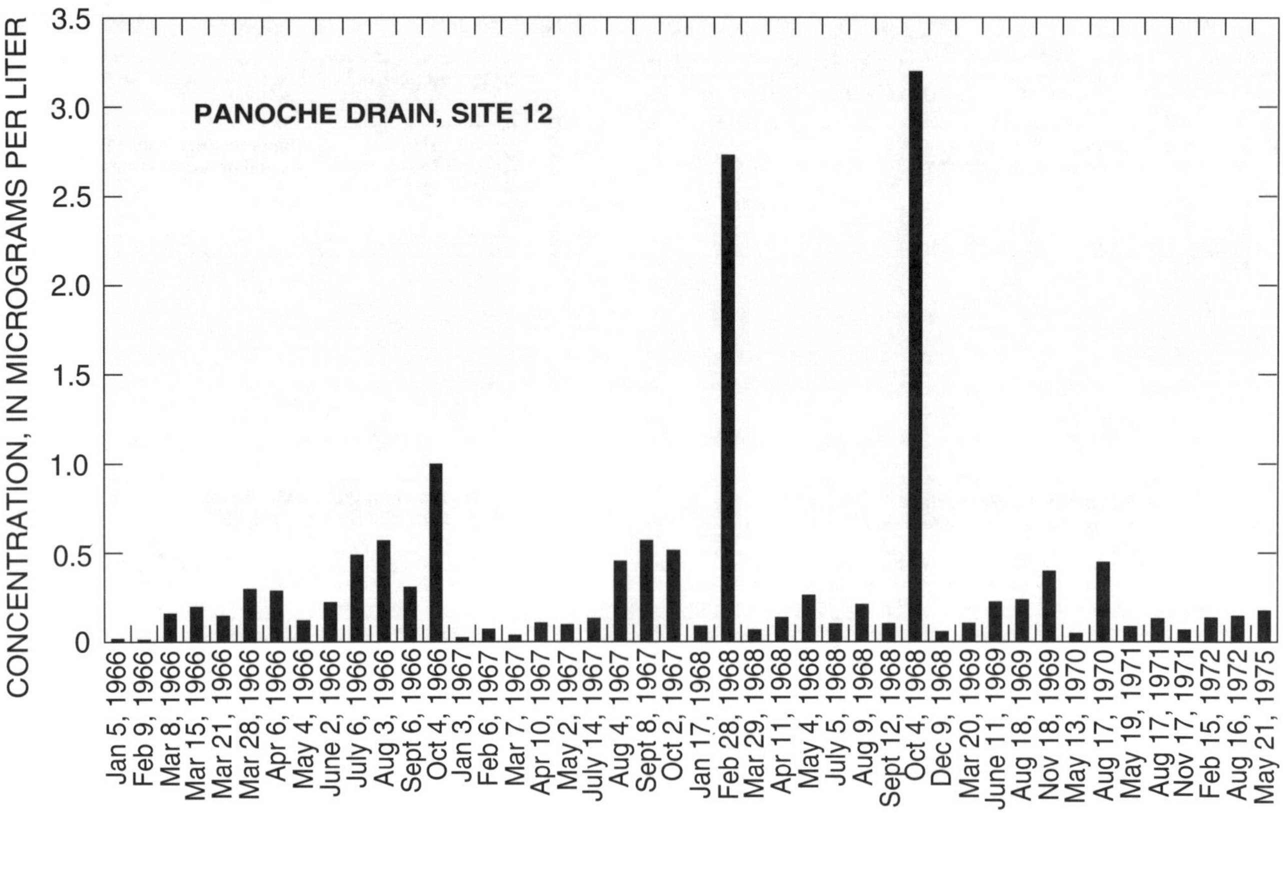

Figure 8. Concentrations of DDT in water samples collected from the Panoche Drain, San Joaquin Basin, California: Valley, west side, alluvial deposits environmental setting—agriculture (field crops). (See figure 6*A* for site location.)

the winter months. The high concentration measured on February 28, 1968, may be attributable to a storm and the introduction of sediment-laden water into the drain. The data for 1971 to 1975 generally indicate lower concentrations than during previous years, but sampling was too infrequent to support an interpretation of decreasing concentrations of DDT in this drain.

Organophosphate pesticides also were detected in tile-drain discharges (figs. 9 and 10). Organophosphate pesticides are used extensively in regions of the

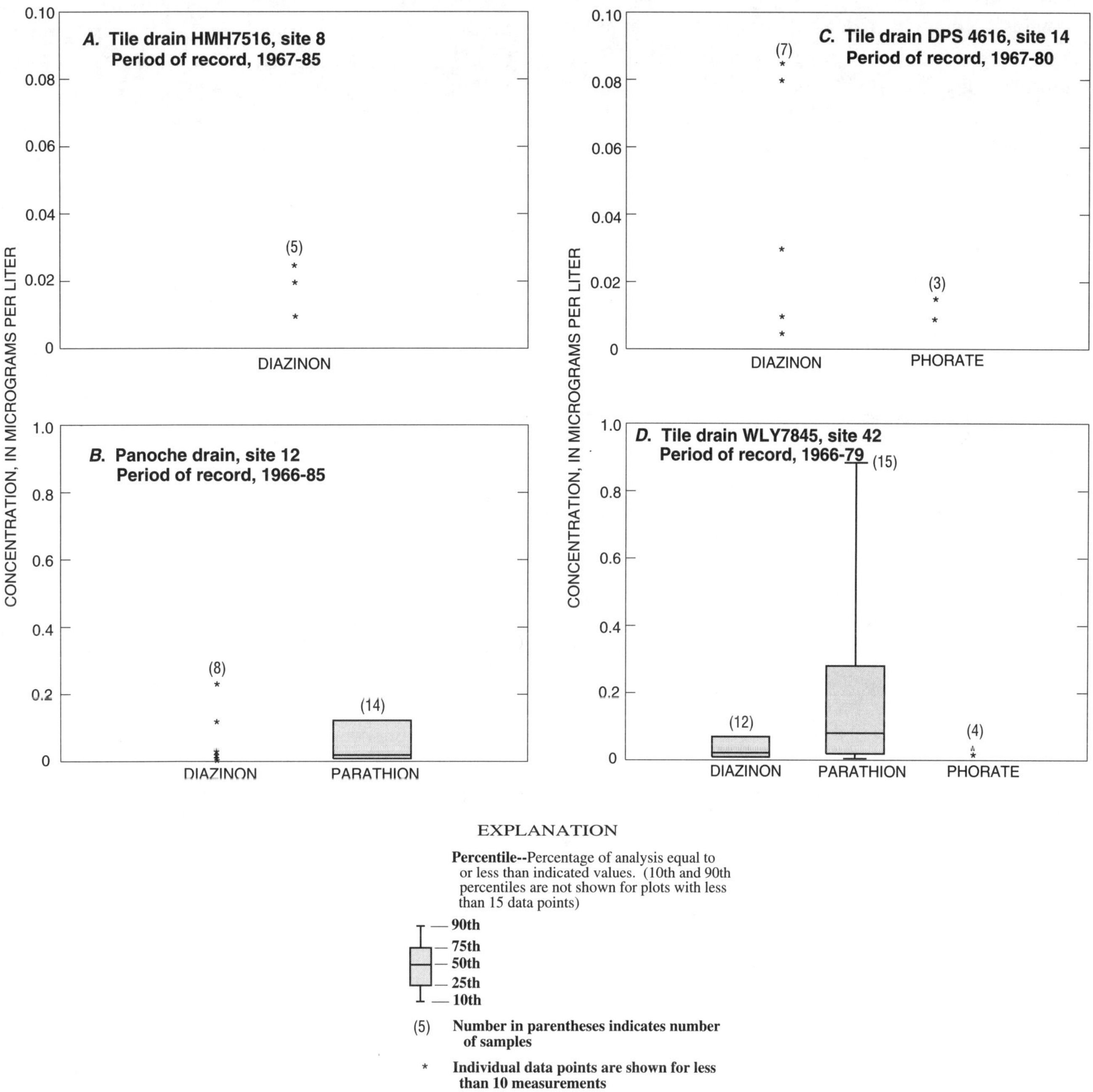

Figure 9. Concentrations of selected organophosphate pesticides in water samples collected from tile drains, San Joaquin Basin, California: Valley, west side, alluvial deposits environmental setting—agriculture (field crops). (Compiled from U.S. Environmental Protection Agency's STORET data base. See figure 6*A* for site locations.)

tile drains, and their presence in tile-drain samples may be due to preferential flow paths of irrigation water. Many of these compounds are applied to vegetation, and losses occur by volatilization from plant surfaces or degradation reactions in the soil horizons. The interpretation of organophosphate pesticides in tile-drain discharge is problematic because these compounds have not been detected in ground water of the study unit (Domagalski and Dubrovsky, 1991, 1992). Further study is needed to determine if these compounds are present in tile-drain discharges and, if so, the mechanism of transport through the soil horizons to the drains.

Triazine herbicides, commonly found in ground water of the study unit, were not detected in tile drainwater by Domagalski and Dubrovsky (1991, 1992), despite extensive sampling with an analytical detection limit of 0.1 µg/L. However, data in the STORET data base indicate that previous researchers detected triazine herbicides in several tile drains within the study unit. These detections need to be verified by new sampling. Data analysis did not distinguish between concentrations of atrazine and simazine but treated the two compounds as one. Median concentrations (fig. 11) ranged from 0.04 µg/L for tile drain CLG7551 (site 35, fig. 6*A*) to 0.63 µg/L at tile drain DPS3465 (site 13, fig. 6*A*). Atrazine and simazine are used mostly as roadside sprays in the area of the tile drains and were detected in shallow observation wells installed near roads (Domagalski and Dubrovsky, 1991, 1992).

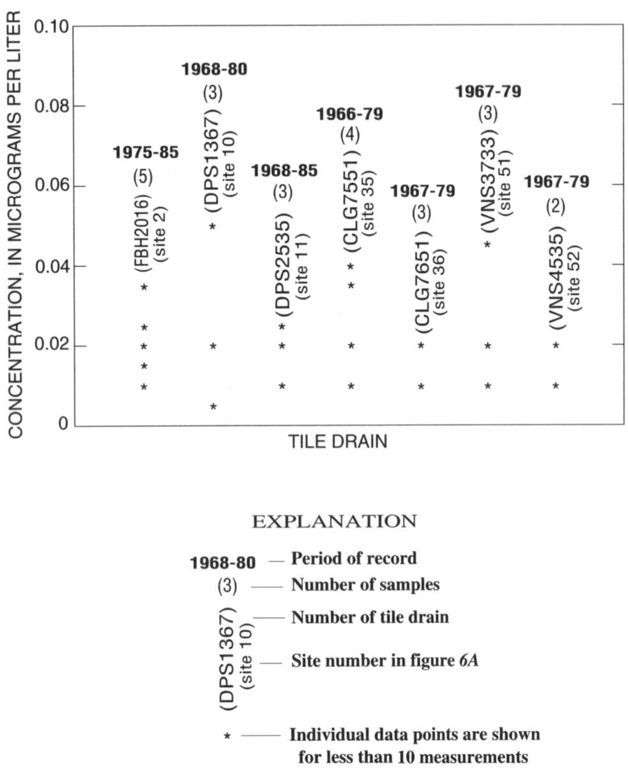

Figure 10. Concentrations of phorate in water samples collected from tile drains, San Joaquin Basin, California: Valley, west side, alluvial deposits environmental setting—agriculture (field crops). (Compiled from U.S. Environmental Protection Agency's STORET data base. See figure 6*A* for site locations.)

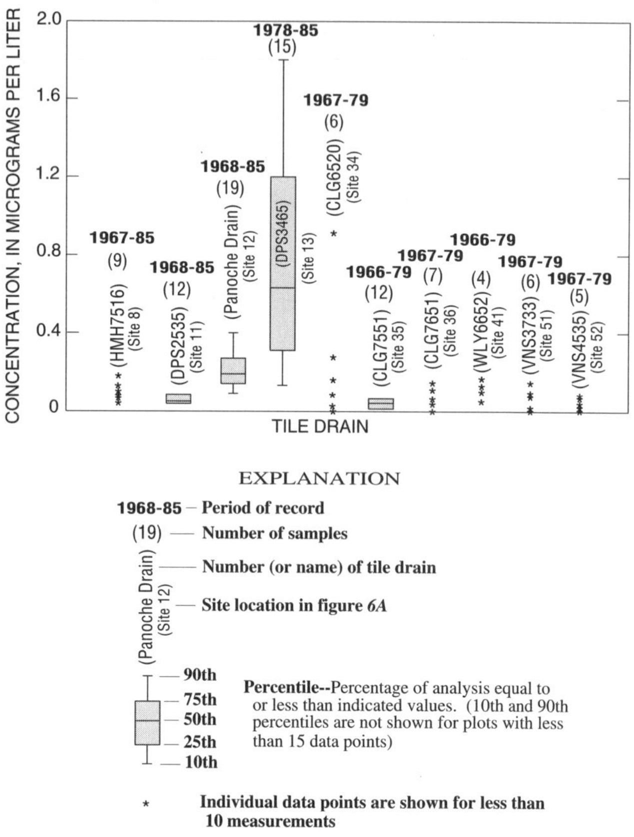

Figure 11. Concentrations of atrazine and simazine in water samples collected from tile drains, San Joaquin Basin, California: Valley, west side, alluvial deposits environmental setting—agriculture (field crops). (Compiled from U.S. Environmental Protection Agency's STORET data base. See figure 6*A* for site locations.)

Data were collected by the California Department of Water Resources at two surface-water sites within this environmental setting. Concentrations of selected organochlorine pesticides, organophosphate pesticides, and triazine herbicides in unfiltered water samples collected from Salt Slough at State Highway 165 near Stevinson (site 21, fig. 6*A*) are shown in figure 12. The highest median concentration of compounds for organochlorine pesticides was 0.035 µg/L for dacthal. Organophosphate pesticides and triazine herbicides detected in water samples from Camp 13 Slough (site 15, fig. 6*A*) are shown in figure 13. Data are not available for organochlorine pesticides in samples from Camp 13 Slough.

San Joaquin Basin Integrator Sites

Integrator sites are stream sampling sites located downstream from large drainage basins that generally contain multiple land uses and environmental settings. Data are available from STORET for two integrator sites, San Joaquin River at Fremont Ford Bridge (site 26, fig. 6*A*) and San Joaquin River near Vernalis (site 50, fig. 6*A*). Sources of pesticides in the San Joaquin River are numerous, but are primarily agricultural. Several organochlorine pesticides were detected in the water column (figs. 14*A* and 15*A*), probably because whole water samples that contained suspended sediment were extracted. In addition to organochlorine pesticides, some organophosphate pesticides were detected at the Fremont Ford Bridge site (fig. 14*B*) and

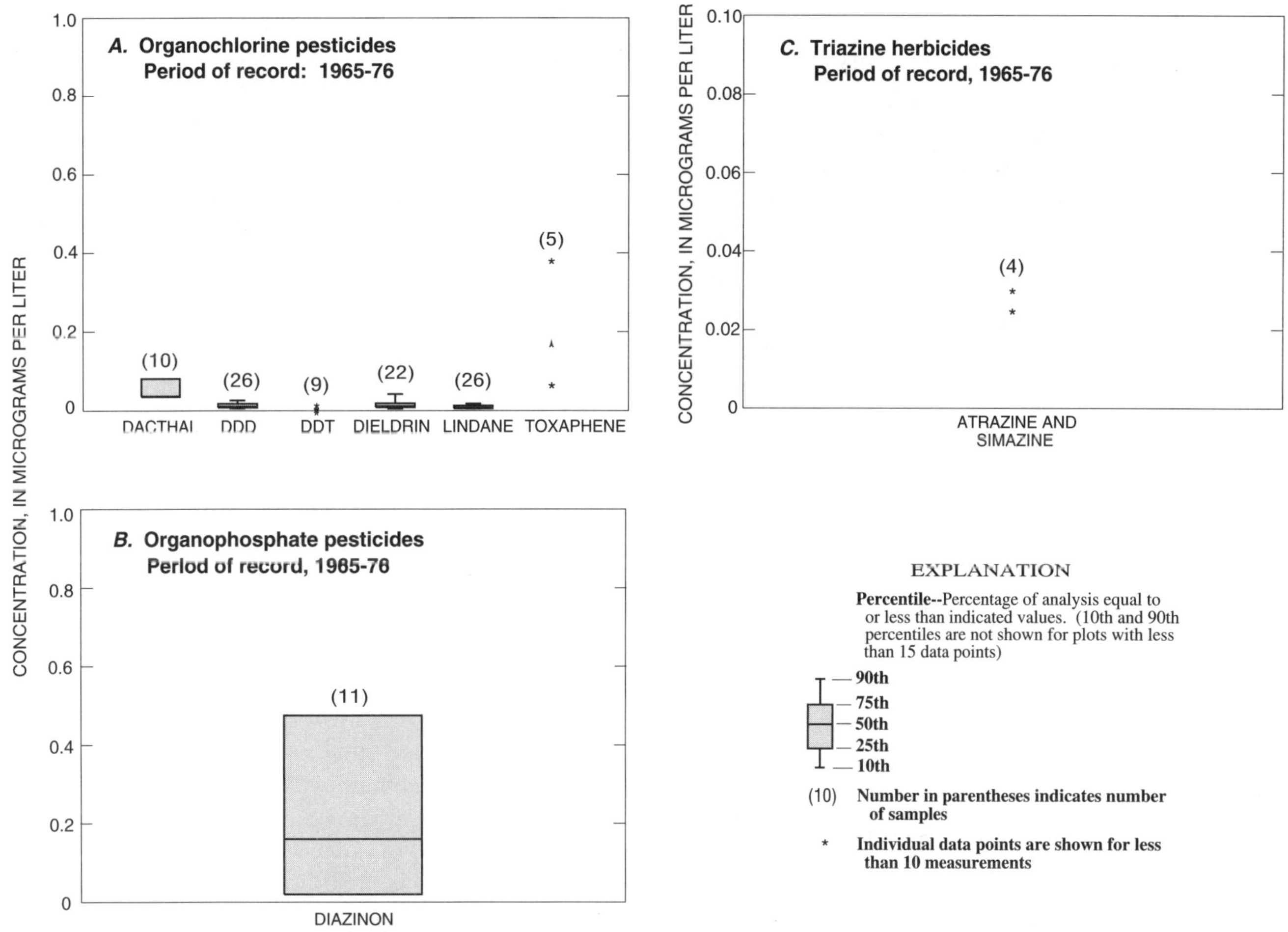

Figure 12. Concentrations of selected pesticides in water samples collected from Salt Slough at State Highway 165 near Stevinson, San Joaquin Basin, California: Valley, west side, alluvial deposits environmental setting—agriculture (field crops). (Site 21, figure 6*A*. Compiled from U.S. Environmental Protection Agency's STORET data base.)

some triazine herbicides were detected at the Vernalis site (fig. 15*B*). The triazine herbicides could have been used in agriculture or as roadway sprays.

Tulare Basin: Valley, West Side Environmental Setting

All pesticide data for the western Tulare Basin in the STORET data base are from tile drainwater. Concentrations of organochlorine pesticides, organophosphate pesticides, and triazine herbicides in water samples collected from selected tile drains are shown in figures 16 through 18. Similar to the San Joaquin Basin, organochlorine pesticides in subsurface tile drainwater are difficult to explain as originating solely through soil leaching. They are more likely due to construction of the tile drains or possibly are transported through the soil with dissolved organic carbon. Surface runoff also may have mixed with tile drainwater prior to sampling. Detections of organophosphate pesticides and triazine herbicides are problematic, as the organophosphates are not commonly detected in ground water. Triazine herbicides, although detected in shallow ground-water observation wells drilled near roads, have not been detected previously in tile drains. These insecticides and herbicides are used extensively in this area, primarily on cotton. The detections in tile drains may be attributable to preferential flow paths of irrigation water to the tile drains.

Tulare Basin: Valley, East Side Environmental Setting

Some organochlorine pesticides were detected in water samples from the Kern, the Kaweah, and the Kings Rivers (fig. 19), and the data are in the STORET data base. Those sites are upstream from the main agricultural region, in the eastern Tulare Basin.

Surface Water—U.S. Geological Survey Data Base, San Joaquin River and Tributaries

Bed Sediment

Bed sediment samples were collected during October 1985 at various locations along the San Joaquin River and a number of tributaries, in zones of fine-sediment deposition (fig. 20) (Gilliom and Clifton, 1990). These samples were analyzed for organochlorine pesticides, and results are given in table 7. In general, the greatest loading of organochlorine pesticides are thought to come from tributary sites west of the San Joaquin River (Gilliom and Clifton, 1990). At some of these sites, especially Orestimba Creek and Ingram Creek, high concentrations can be attributable to fine-grained, easily erodible soils and previous high use of DDT. Organochlorine pesticides tend to concentrate in fine-grained sediment. Salt Slough has a predominance of fine-grained sediment and, therefore, is a potential source of high organochlorine pesticide loading. With the possible exception of the Merced River, drainages east of the San Joaquin River are not thought to contribute significant loading of organochlorine pesticides to the river. However, not all potential sources of organochlorine pesticides were monitored by Gilliom and Clifton (1990).

The San Joaquin River near Vernalis is the most downstream sampling site for the San Joaquin–Tulare Basins study unit. Concentrations of organochlorine pesticides in bed sediments at the San Joaquin River near Vernalis station are among the highest reported for major rivers of the United States (Gilliom and Clifton, 1990).

Water Column

Many of the bed sediment sampling sites also were sampled for pesticides in the water column. Synoptic samplings (short-term investigations of specific water-quality conditions during selected seasonal or hydrologic periods) were done during the low-flow season in 1985 and 1987. The results of synoptic sampling in the water column are presented in table 8; the site locations are shown in figure 21.

The water column also was sampled at the San Joaquin River near Vernalis as part of the National Stream Quality Accounting Network during 1985-87. Concentrations of organochlorine pesticides detected in unfiltered water samples from the San Joaquin River at Patterson Bridge near Patterson (site 37, fig. 21) and the San Joaquin River near Vernalis (site 50, fig. 21) are shown in figure 22. The concentrations of organochlorine pesticides tend to be low. This was expected because the organochlorine pesticides are mostly in suspended sediment. The highest concentration was 0.07 µg/L for endosulfan at the San Joaquin River near Patterson.

The USGS Toxic Substances Hydrology Program also sampled for pesticides near Vernalis. Samples were collected daily, and 2-day composites were analyzed for selected pesticides including diazinon, parathion, carbofuran, and triazine herbicides. The analyses were completed at the USGS laboratory in Sacramento, California, by solid-phase extraction and

Table 7. Concentrations of selected organochlorine pesticides in bed sediment samples, San Joaquin Basin, California

[See figure 20 for site locations. Concentrations are in microgram per kilogram. DDD, 1,1-dichloro-2,2-bis (4-chlorophenyl) ethane; DDE, 1,1-dichloro-2,2-bis (4-chlorophenyl) ethylene; DDT, 1,1,1-trichloro-2,2-bis (4-chlorophenyl) ethane. <, less than]

Site no.	Station name	Aldrin	Chlordane	DDD	DDE	DDT	Dieldrin	Endosulfan	Toxaphene
	Valley, west side, alluvial deposits, and agriculture (field crops)								
20	Santa Fe Canal at Gun Club Road	<0.1	<0.1	<0.1	0.4	<0.1	<0.1	<0.1	<10
21	Salt Slough at State Highway 165 near Stevinson	<.1	<1.0	7.2	12	<.1	.9	<.1	<10
23	Mud Slough near Gustine	<.1	<1.0	.7	1.3	<.1	.1	<.1	<10
27	Newman Wasteway near State Highway 33	<.1	<1.0	21	130	<.1	2.0	<.1	<10
	Valley, west side, alluvial deposits, and agriculture (row crops)								
31	Orestimba Creek at State Highway 33	0.9	<1.0	180	430	55	6.8	15	<10
44	Ingram Creek at River Road	<.1	<1.0	260	250	420	4.9	87	<10
45	Hospital Creek at River Road	<.1	<1.0	58	170	60	8.9	<.1	<10
	Valley, west side, alluvial deposits, and agriculture (orchards)								
39	Del Puerto Creek at State Highway 33	<0.1	1.0	29	73	<0.1	1.4	<0.1	250
	Basin deposits and agriculture (field crops)								
25	Los Banos Creek at State Highway 140	<0.1	<1.0	0.2	0.6	<0.1	<0.1	<0.1	<10
	Valley, east side, alluvial deposits, and agriculture (field crops)								
1	Fresno Slough near State Highway 180	<0.1	<1.0	1.6	6.4	<0.1	<0.1	<0.1	<10
22	Bear Creek near Stevinson	<.1	<1.0	<.1	.1	<.1	<.1	<.1	<10
	Integrator sites								
3	San Joaquin River below Mendota Pool at Bass Avenue	<0.1	<1.0	<0.1	0.9	<0.1	<0.1	<0.1	<10
9	San Joaquin River at Firebaugh Bridge	<.1	1.0	.9	3.5	.6	.1	<.1	<10
16	San Joaquin River near Dos Palos	<.1	<1.0	.3	.4	.2	<.1	<.1	<10
17	San Joaquin River at Washington Road Bridge	<.1	<1.0	1.6	3.8	<.1	<.1	<.1	<10
18	San Joaquin River at Turner Island at Erreca Road	<.1	<1.0	11	58	<.1	4.4	2.6	<10
19	Mariposa Bypass-Slough	<.1	<1.0	1.6	3.9	<.1	.7	.8	<10
24	San Joaquin River near Stevinson	<.1	1.0	.3	1.2	<.1	<.1	<.1	<10
26	San Joaquin River at Fremont Ford Bridge	<.1	<1.0	.3	.7	<.1	<.1	<.1	<10
29	Merced River near Stevinson	<.1	<1.0	3.9	4.2	49	.3	<.1	<10
43	Tuolumne River at Modesto	<.1	<1.0	<.1	.1	<.1	<.1	<.1	<10
48	Stanislaus River at Ripon	<.1	2.0	1.3	2.4	1.1	.1	<.1	<10
50	San Joaquin River near Vernalis	<.1	3.0	3.2	7.1	1.3	1.0	<.1	<10

gas chromatography/mass spectrometry. The methodology is described in a recent report by Crepeau and others (1994).

Concentrations of organophosphate pesticides detected in water samples from the San Joaquin River at Patterson Bridge near Newman (site 28, fig. 21), near Patterson, and near Vernalis are shown in figure 23. Diazinon is one of several organphosphate pesticides, in addition to chlorpyrifos, methidathion, and parathion, sprayed on dormant orchard and other stone-fruit orchards. Diazinon also is applied to alfalfa fields. The approximate amounts of diazinon applied to dormant orchards and alfalfa fields throughout the study unit during 1990 are shown in figures 24 and 25. The greatest amounts of diazinon are applied on dormant almond orchards in January and on alfalfa fields in March (figs. 26 and 27*B*). Concentrations of diazinon in water samples collected from the San Joaquin River near Vernalis tend to increase at the end of January and the beginning of February and again in March (fig. 28*B*). These increases can probably be attributed to the increased spraying and to the rainstorms that can mobilize the diazinon to the San Joaquin River.

Carbofuran concentrations in water samples collected at the San Joaquin River near Vernalis are shown in fig. 28*A*. Carbofuran, one of several pesticides used on alfalfa fields, is not sprayed on dormant orchards and was not detected during January and February. The detections of carbofuran during March probably resulted from transport from the sprayed alfalfa fields to the river during March rainstorms. The locations of alfalfa fields where carbofuran was applied and the amounts applied in 1990 are shown in figure 29. The amounts applied monthly in 1990 are shown in figure 27*A*.

Water samples were collected at the San Joaquin River near Vernalis during the week of February 10, 1992, the first significant rainstorm of that winter. Analysis of these samples for selected pesticides, including carbaryl, carbofuran, dacthal, diazinon, malathion, methidathion, parathion, and triazine herbicides was done by solid-phase extraction and gas chromatography/mass spectrometry. Pesticides detected

Table 8. Results of synoptic sampling in the water column in the San Joaquin River and selected tributaries, San Joaquin Valley, California

[See figure 21 for site locations. µg/L, microgram per liter]

Site no.	Station name	Compound	Concentration (µg/L)
	Autumn 1985		
6	Firebaugh Wasteway at Helm Canal Road	Ethion	0.01
16	San Joaquin River near Dos Palos	Chloroform	13
		Trichloroethylene	6.5
17	San Joaquin River at Washington Road Bridge	Parathion	.15
18	San Joaquin River at Turner Island at Erreca Road	Parathion	.01
25	Los Banos Creek at State Highway 140	Dicamba	.01
27	Newman Wasteway near State Highway 33	2,4-D	.09
31	Orestimba Creek at State Highway 33	Diazinon	.07
39	Del Puerto Creek at State Highway 33	Dicamba	.01
45	Hospital Creek at River Road	Dicamba	.01
	Late summer 1987		
28	San Joaquin River near Newman	Dicamba	0.01
		Cyanazine	.1
		Simazine	.1
		Methomyl	2.3
37	San Joaquin River at Patterson Bridge, near Patterson	Ethion	.01
		2,4-D	.01
		Metholachlor	.1

include carbaryl, dacthal, diazinon, malathion, methidathion, and simazine (fig. 30). The concentrations of diazinon were considerably higher than those detected during the dormant-orchard spray season in 1991 (fig. 24*B*) and can be attributed to higher rainfall during the dormant season of 1992 that mobilized pesticides from the orchards to the rivers. Methidathion also is used as a spray for dormant almond orchards. Simazine is

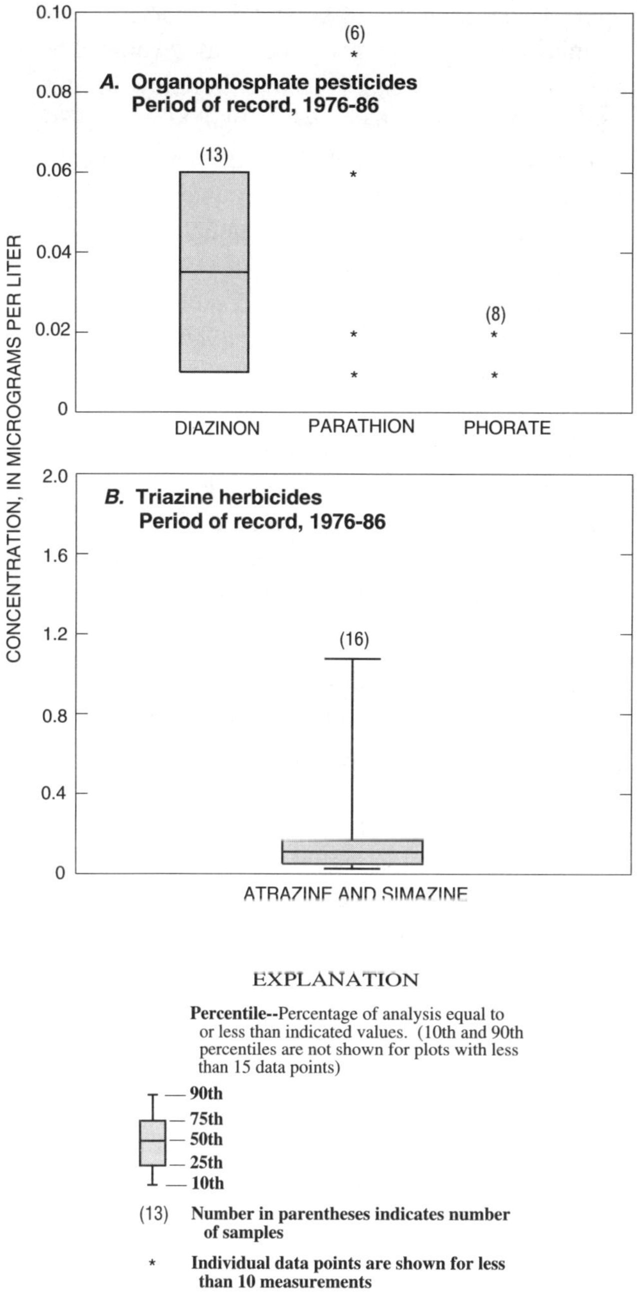

Figure 13. Concentrations of selected pesticides in water samples collected from Camp 13 Slough, San Joaquin Basin, California: Valley, west side, alluvial deposits environmental setting—agriculture (field crops). (Site 15, figure 6*A*. Compiled from U.S. Environmental Protection Agency's STORET data base.)

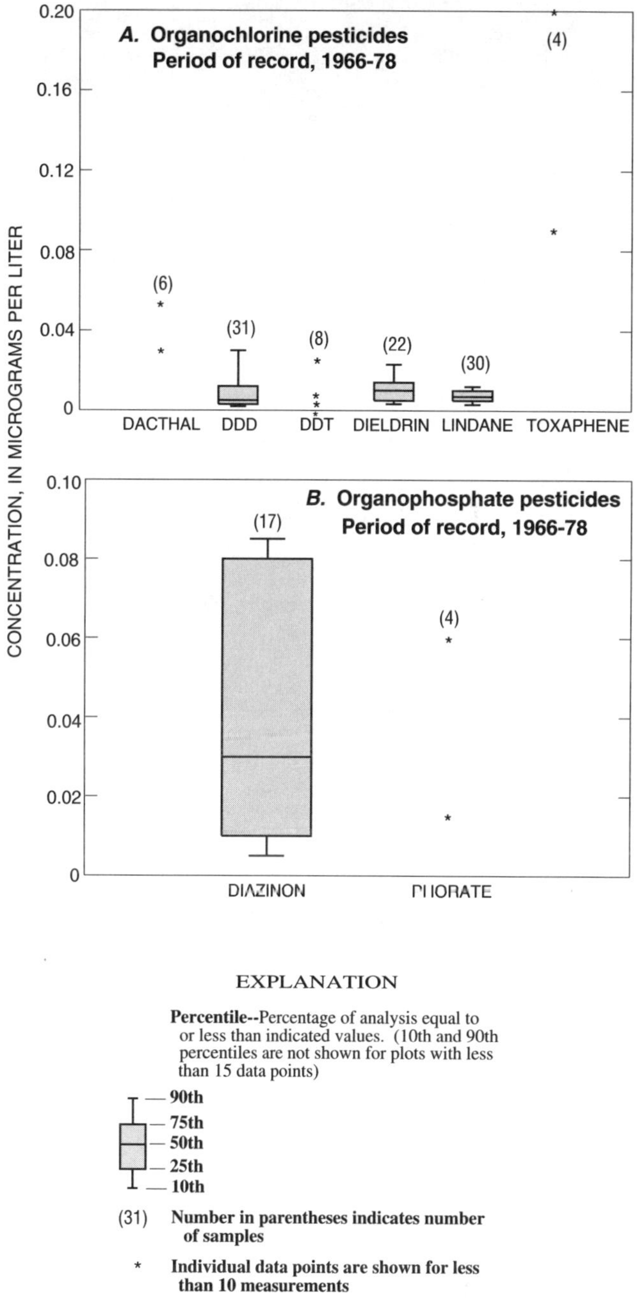

Figure 14. Concentrations of selected pesticides in water samples collected from the San Joaquin River at Fremont Ford Bridge integrator site, San Joaquin Basin, California. (Site 26, figure 6*A*. Compiled from U.S. Environmental Protection Agency's STORET data base.)

used in agriculture and as a roadside spray; dacthal is used on a variety of row crops, including onions and tomatoes. The recorded use of carbaryl and malathion

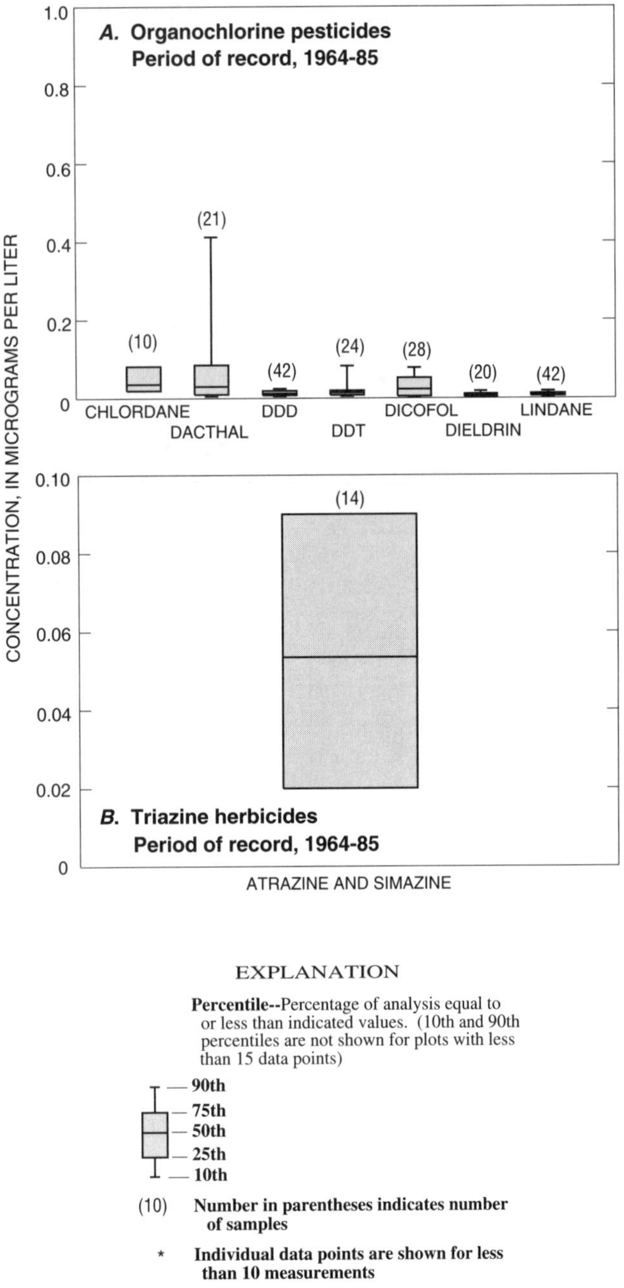

Figure 15. Concentrations of selected pesticides in water samples collected from the San Joaquin River near Vernalis integrator site, San Joaquin Basin, California. (Site 50, figure 6*A*. Compiled from U.S. Environmental Protection Agency's STORET data base.)

during February 1992 indicates only limited application, and the measured concentrations were low.

The concentration of diazinon in the San Joaquin River near Vernalis exceeded 0.35 µg/L during the February 1992 storm. That concentration is toxic to *Ceriodaphnia dubia*, an organism used in the three-species bioassay of potentially toxic effluents (Amato and others, 1992), which was developed by the U.S. Environmental Protection Agency (Mount and Anderson-Carnahan, 1988, 1989; Mount, 1989). This bioassay is used to evaluate the management or regulation of point- or nonpoint-source contamination of surface water. In contrast, all samples collected during the dormant spray period of 1991, when rainfall was lower, are less than the threshold of 0.35 µg/L. Pesticide concentrations were relatively low because there was little rainfall in January 1991, and therefore less runoff from the orchards. Diazinon has no current regulatory standards for the protection of aquatic life. Simazine concentrations exceeded those of diazinon during the February 1992 storm, but are well below the U.S. Environmental Protection Agency drinking-water standards of 4 µg/L for simazine.

Surface Water—California Regional Water Quality Control Board Data Base, San Joaquin River and Tributaries

The California Regional Water Quality Control Board, in cooperation with the USGS, collected water samples during 1991 and 1992 at sites on the San Joaquin River and adjacent agricultural drains and tributaries (fig. 31). The purpose of the study was to determine if water samples were toxic to three test organisms—fathead minnows (*Pimephales promelas*), a cladoceran zooplankter (*Ceriodaphnia dubia)*, and a green alga (*Selenastrum capricornutum)*. In previous studies (Foe and Connor, 1991), toxicity was observed most frequently in the zooplankter, although some samples also were toxic to the fathead minnow. All toxic samples, as well as some nontoxic samples, were analyzed by the USGS, using the methods described in a report by Wershaw and others (1987) to determine if organophosphate or carbamate pesticides were present in sufficient concentration to explain the toxicity. Organophosphate and carbamate pesticides were selected for analysis because they are acetylcholinesterase inhibitors and, therefore, are potentially toxic to aquatic organisms. Although the quality assurance of the analytical laboratory is known, the California Regional Water Quality Control Board used nonstandard techniques to collect water samples. Most water samples were point samples collected off the bank or by wading, where appropriate, by dipping a bottle into the stream. The toxicity testing protocols are

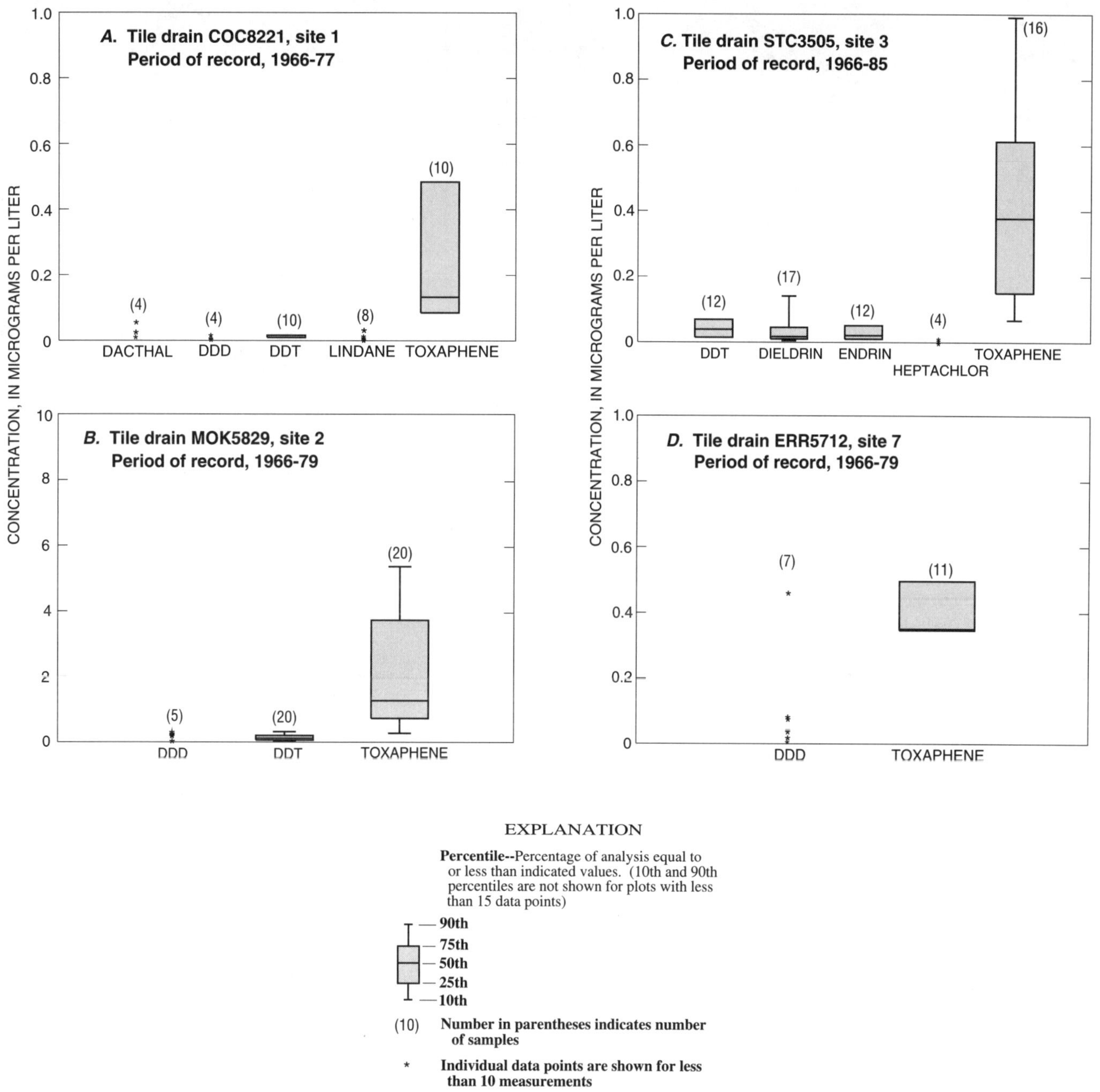

Figure 16. Concentrations of organochlorine pesticides in water samples collected from selected tile drains, Tulare Basin, California: Valley, west side environmental setting. (Compiled from U.S. Environmental Protection Agency's STORET data base. See figure 6*B* for site locations.)

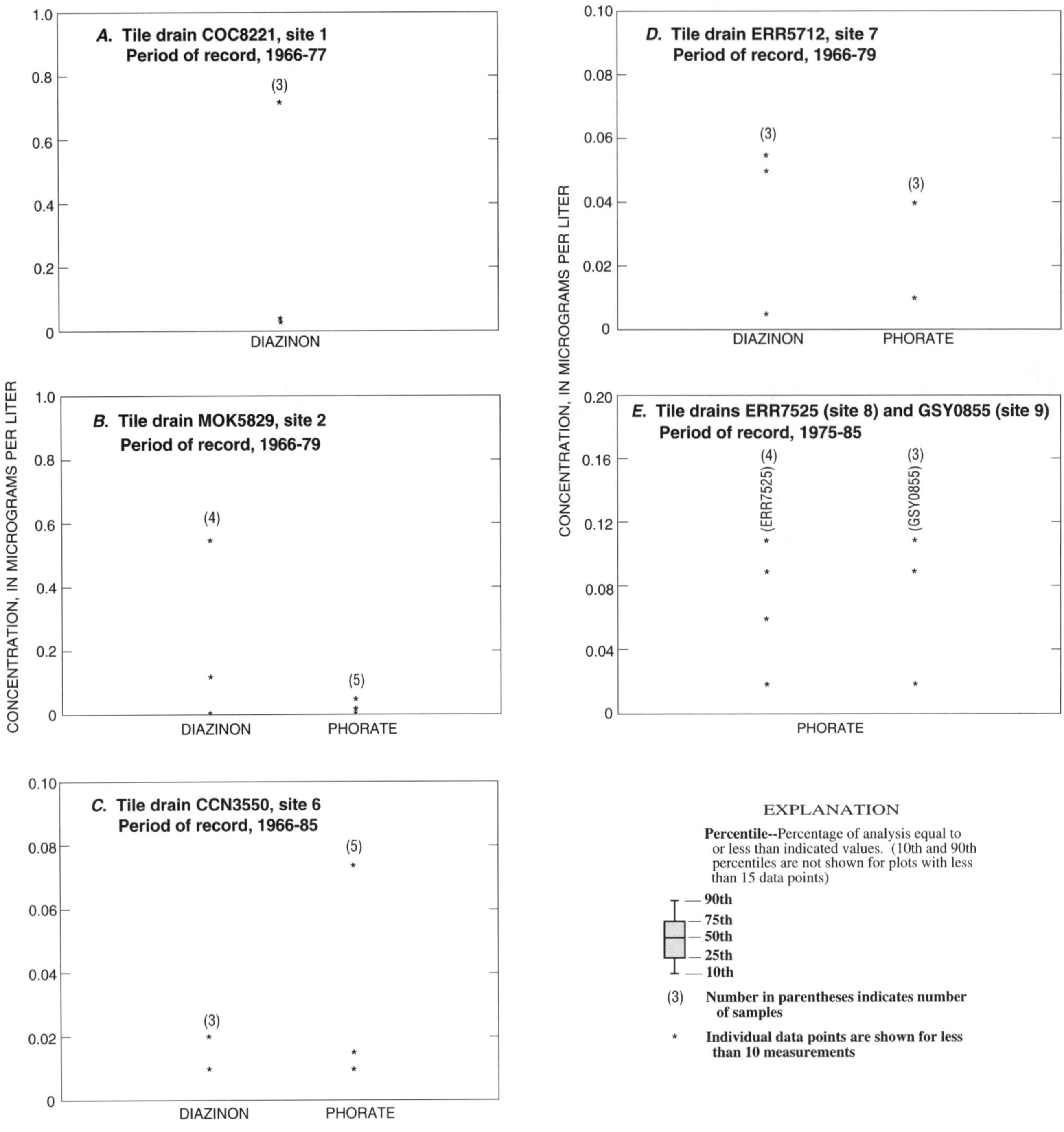

Figure 17. Concentrations of organophosphate pesticides in water samples collected from tile drains, Tulare Basin, California: Valley, west side environmental setting. (Compiled from U.S. Environmental Protection Agency's STORET data base. See figure 6*B* for site locations.)

those specified by the U.S. Environmental Protection Agency (Mount and Anderson-Carnahan, 1988, 1989; Mount, 1989).

San Joaquin Basin: Valley, West Side, Alluvial Deposits Environmental Setting—Agriculture (Row Crops, Orchards, and Field Crops)

Chlorpyrifos and diazinon were the most frequently detected organophosphate pesticides in row crops. The median concentrations ranged from 0.02 to 0.06 µg/L; however, concentrations can vary considerably (fig. 32*A-C*). Parathion was detected at the Spanish Grant Drain (site 32, fig. 31).

Chlorpyrifos, diazinon, and parathion were the most frequently detected organophosphate pesticides

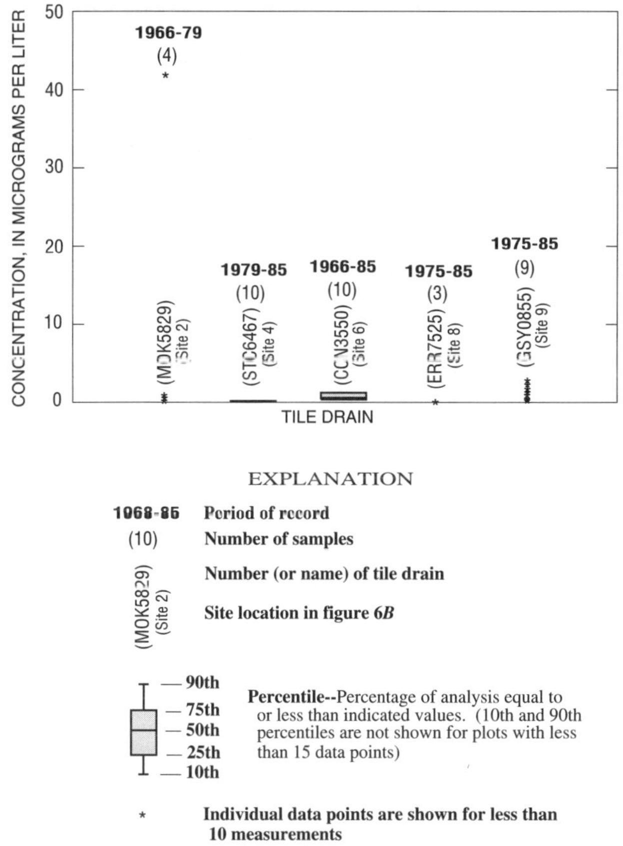

Figure 18. Concentrations of atrazine and simazine in water samples collected from tile drains, Tulare Basin, California: Valley, west side environmental setting. (Compiled from U.S. Environmental Protection Agency's STORET data base. See figure 6*B* for site locations.)

in orchards. The median concentrations ranged from 0.01 to 0.04 µg/L (fig. 33). The most variation in concentration is for diazinon. In contrast, the least variation in concentration is for chlorpyrifos.

Chlorpyrifos, diazinon, and parathion were the most frequently detected organophosphate pesticides in field crops (fig. 34). The median concentration of diazinon for samples collected at Salt Slough was 0.03 µg/L. However, concentrations can vary considerably.

San Joaquin Basin: Valley, East Side, Alluvial Deposits Environmental Setting—Agriculture (Orchards and Field Crops)

The median concentrations of chlorpyrifos and diazinon were 0.01 and 0.04 µg/L, respectively. The greatest variation in concentration is for diazinon (fig. 35).

Concentrations of chlorpyrifos, diazinon, and parathion at Turlock Irrigation District drain lateral 6, in the field crops areas, are shown in figure 36*A*. The median concentrations of chlorpyrifos and diazinon at the Turlock Irrigation District drain lateral 3 were 0.04 and 0.09 µg/L, respectively. Concentrations of both compounds vary considerably at this site (fig. 36*B*).

San Joaquin Basin Integrator Sites

Chlorpyrifos, diazinon, and parathion were the most frequently detected organophosphate pesticides, but fonofos was detected at some sites. The detection limit was 0.01 µg/L. Median concentrations of chlorpyrifos and diazinon range from 0.01 to 0.05 µg/L. The concentrations of diazinon indicate more variation than those of chlorpyrifos (fig. 37*A-F*). All samples were analyzed for carbamate pesticides also, but there were only a few detections, probably due to the relatively high detection limit of 0.5 µg/L (Wershaw and others, 1987).

Chlorpyrifos was detected frequently but at lower concentrations than diazinon. Applications of diazinon and chlorpyrifos are high throughout the San Joaquin Valley. However, application of chlorpyrifos during the dormant-orchard season is less than diazinon, and the concentrations should be lower during this period. The application of chlorpyrifos is high during the alfalfa spraying season, but rainfall, and therefore runoff, is less at that time. Chlorpyrifos has lower solubility than diazinon; thus, partitioning of chlorpyrifos from water to sediment occurs and the chlorpyrifos concentration in runoff is less.

At most stations, the data indicate considerable variation in the concentrations of some organophosphate pesticides (figs. 32-37). The source of that variability can be examined by plotting the individual data for a site. Diazinon data for Orestimba Creek at State Highway 33 (site 31, fig. 31) are shown in figure 38. Concentrations were higher than the median of 0.02 μg/L during four periods; March 1991, February 1992, March 1992, and May 1992. Diazinon is applied to alfalfa fields during March. The March 1991 increases in diazinon concentration in the San Joaquin River can be attributed to a series of storms that occurred throughout the month. The increase during February 1992 can be attributed to the application of diazinon as a spray on dormant almond orchards and to a rainstorm that occurred on or about February 10, 1992. The concentrations decreased toward the end of February and then increased on March 24, 1992. That increase can be attributed to the spraying of diazinon on alfalfa fields and to a storm just prior to March 24. The increasing concentrations during May 1992 are the highest for this data set. The explanation of this increase is more problematic because there was no rainfall during this period, and the mechanism of transport from the field to the river cannot be assessed at this time. However, diazinon probably was sprayed on almond and other orchards, and the increase possibly was due to diazinon runoff into irrigation water.

A possible relation between measured pesticide concentrations and the percentage of orchard acreage

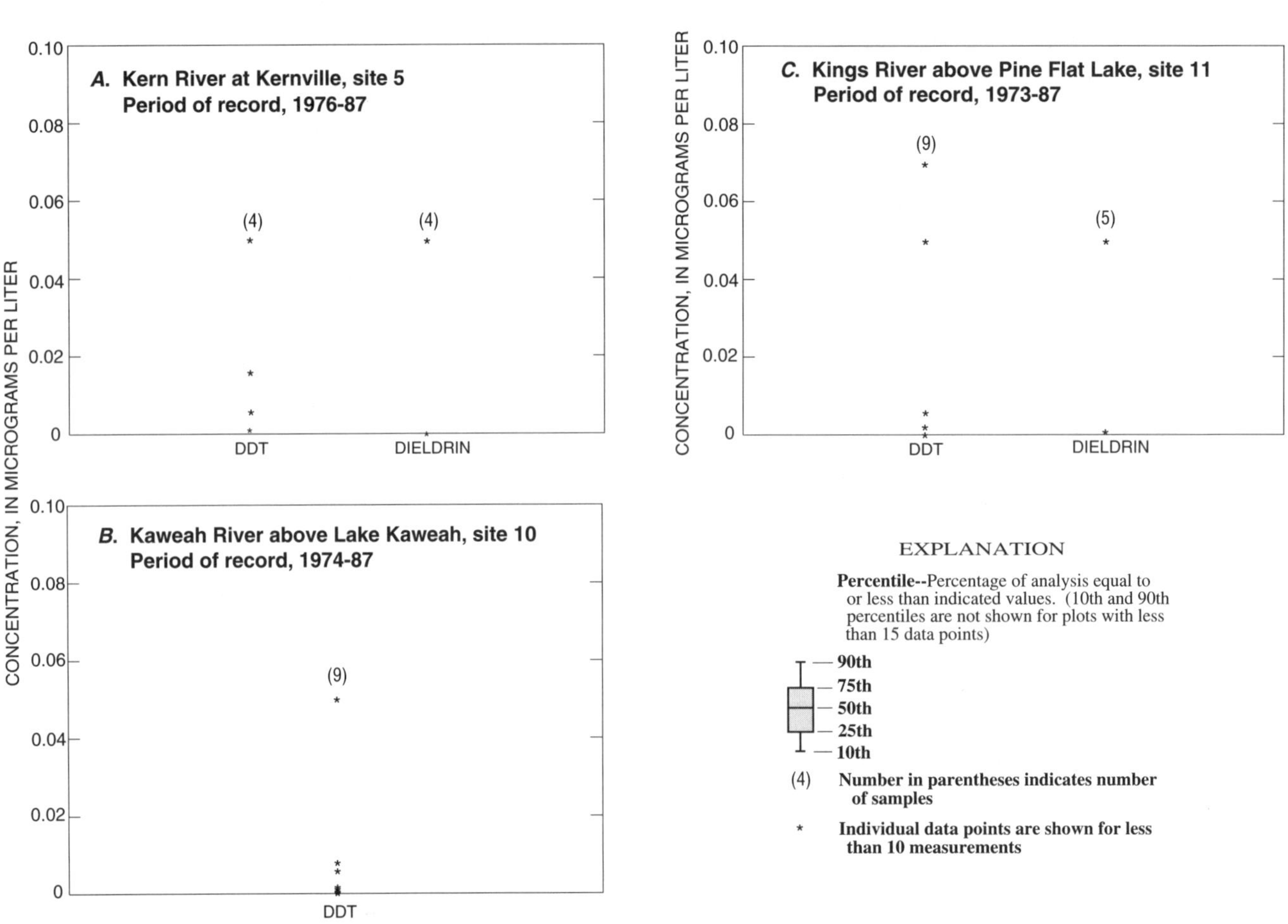

Figure 19. Concentrations of organochlorine pesticides in water samples collected from the Kern, the Kaweah, and the Kings Rivers, Tulare Basin, California: Valley, east side environmental setting. (Compiled from U.S. Environmental Protection Agency's STORET data base. See figure 6*B* for site locations.)

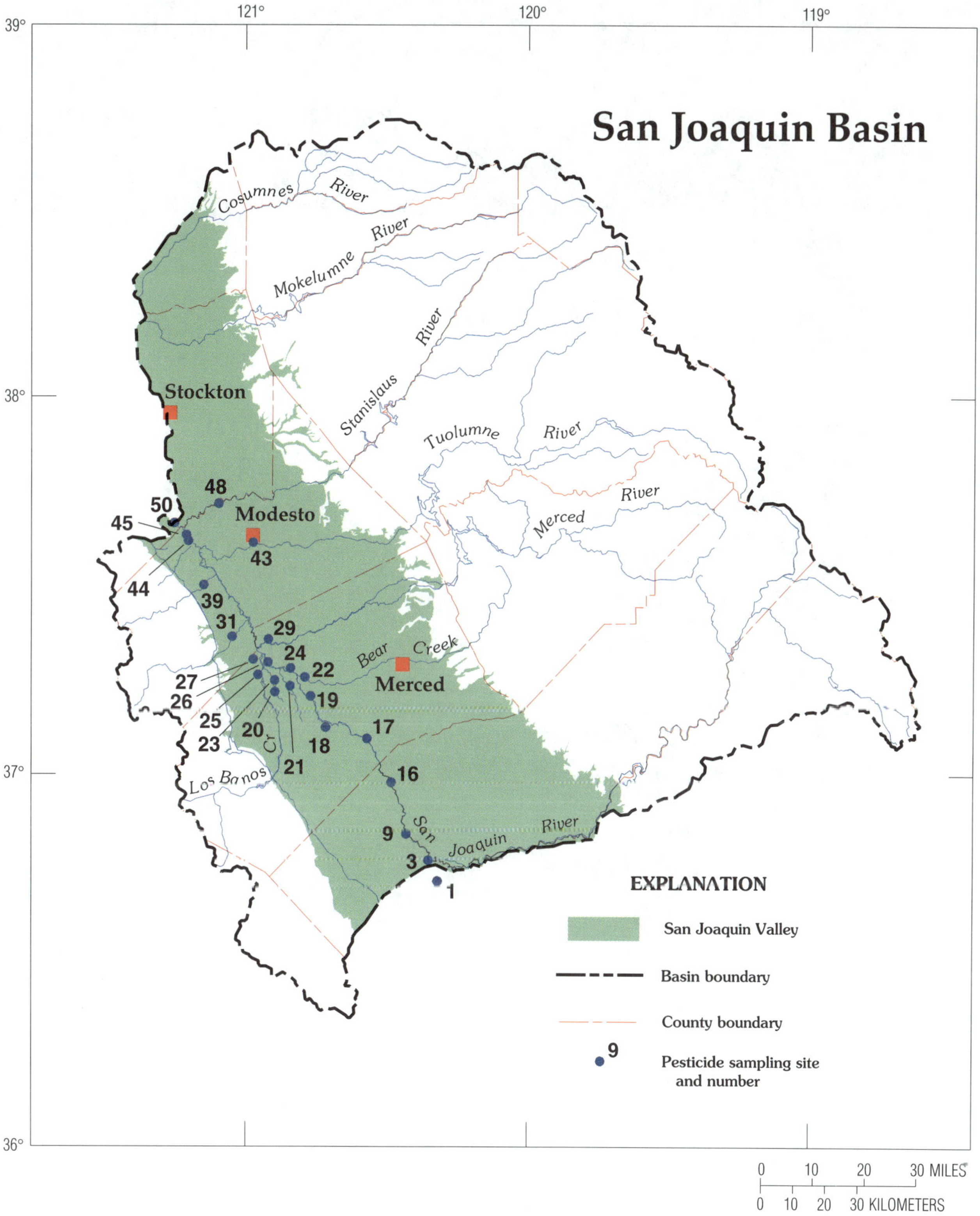

Figure 20. Locations of sites sampled for organochlorine pesticides in bed sediment along the San Joaquin River, San Joaquin Basin, California. (Compiled from U.S. Geological Survey data base. See table 6 for station numbers and descriptions.)

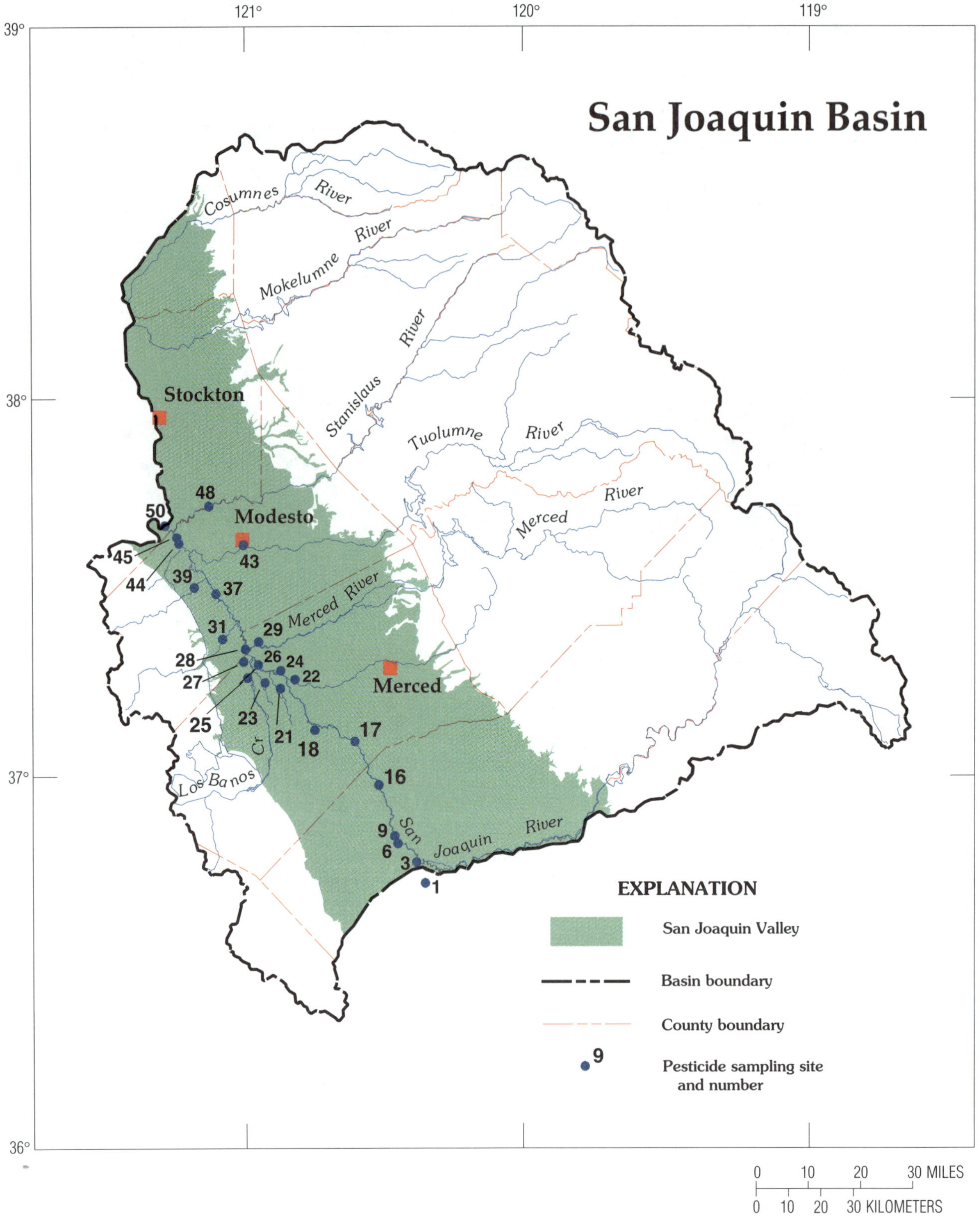

Figure 21. Locations of sites sampled for pesticides in the water column of the San Joaquin River, San Joaquin Basin, California. (Compiled from U.S. Geological Survey data base. See table 6 for station numbers and descriptions.)

in a drainage area can be seen by comparing the data for diazinon at three stream sites in the western San Joaquin Valley. The concentrations of diazinon for Orestimba Creek at State Highway 33 (site 31, fig. 31), Spanish Grant Drain (site 32, fig. 31), and Del

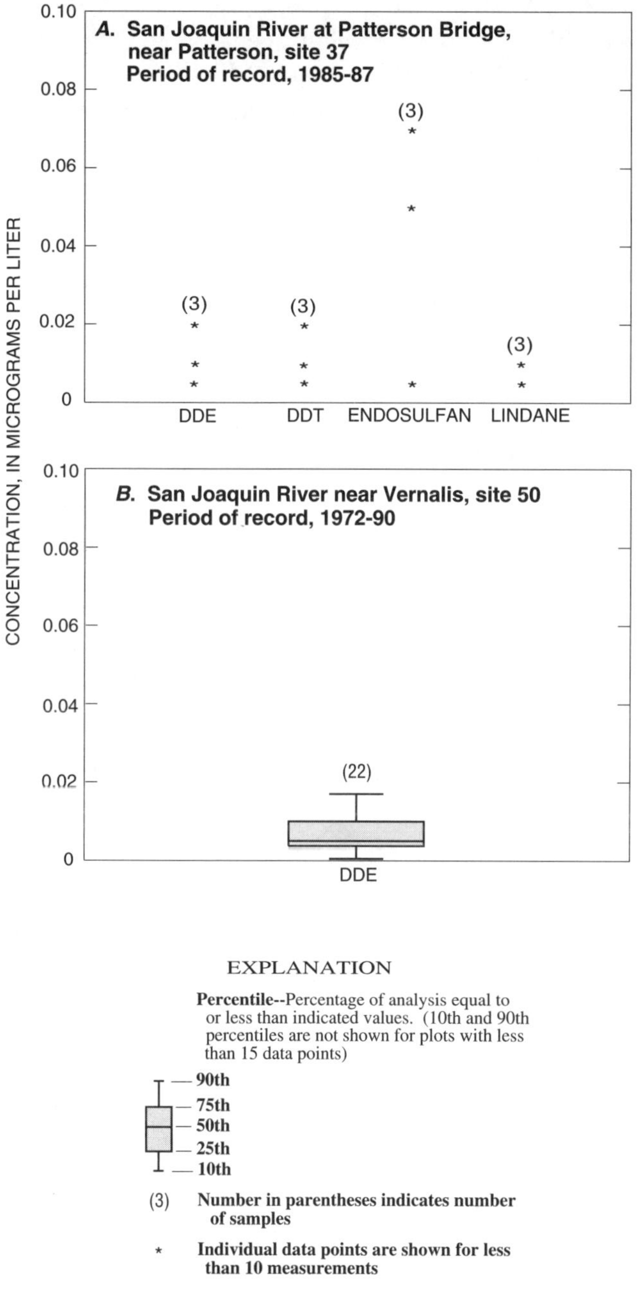

Figure 22. Concentrations of organochlorine pesticides in water samples collected from the San Joaquin River near Patterson and near Vernalis, San Joaquin Basin, California. (Compiled from U.S. Geological Survey data base. See figure 21 for site locations.)

Puerto Creek at State Highway 33 (site 39, fig. 31) are shown in figure 39. Orchard acreage is 40 percent of land use in the Del Puerto Creek drainage, 25 percent in the Orestimba Creek drainage, and 12 percent in the Spanish Grant Drain drainage. During the dormant-orchard spray season, diazinon concentrations in water samples from Orestimba and Del Puerto Creeks generally were higher than those in samples from the Spanish Grant Drain, possibly because of the difference in the relative amount of orchard acreage.

Individual diazinon concentrations at Turlock Irrigation District drain lateral 5 are shown in figure 40. This site is in the east valley, orchards environmental setting. The pattern of concentrations has some similarities to that of Orestimba Creek (fig. 39*A*), as well as some differences. The two graphs are not directly comparable because the two sites were not always sampled on the same day.

Diazinon concentrations greater than the median concentration of 0.04 μg/L were recorded at Turlock Irrigation District drain lateral 5 during late January through February 1992 (fig. 40). Some of the higher concentrations cannot be attributed to rainfall, especially the samples collected on January 30 and February 3. Those concentrations probably are the result of diazinon spraying on almond and other stone-fruit orchards, and the transport may have resulted from irrigation drainage. Concentrations measured in samples collected between February 10 and 24 probably are attributable to rainfall. The soils of the eastern alluvial fans of the San Joaquin Valley, where many almond orchards are located, are coarse grained relative to those of the western alluvial fans. Therefore, infiltration rates are thought to be higher, and runoff of irrigation water should not occur. Nevertheless, the detections of diazinon at Turlock Irrigation District drain lateral 5 prior to the February 10, 1992, storm cannot be attributed to rainfall.

The effect of rainfall on the concentrations of organophosphate pesticides, such as chlorpyrifos and diazinon, can be seen by examining data sets of Lagrangian and synoptic samplings. Lagrangian samplings are designed to measure concentration changes in a moving parcel of water. The California Regional Water Quality Control Board participated in a cooperative Lagrangian sampling with the California Department of Pesticide Regulation on January 29-30, 1992, to assess sources of pesticides or dilution water. Two synoptic studies by the California Regional Water Quality Control Board sampled the perennial part of the San Joaquin River. Samples were submitted to the USGS laboratory in Arvada, Colorado, for analysis.

Concentrations of chlorpyrifos and diazinon, the only two organophosphate pesticides detected in the synoptic studies, are shown in table 9. At most sites, concentrations of diazinon on January 29-30 were only slightly greater than the detection limit of 0.01 µg/L. During Lagrangian sampling, the only high concentration of diazinon was at Turlock Irrigation District drain lateral 5. That high concentration was confirmed by independent analyses by the USGS and the California Department of Pesticide Regulation laboratories. The Lagrangian sampling showed only minor amounts of diazinon in the San Joaquin River prior to the rainstorm of February 10, 1992.

An intense storm occurred during the week of February 10, 1992. The California Regional Water Qualtiy Control Board sampled selected sites on the river and tributaries on February 10-12, 1992, and February 17-19, 1992. The results of those analyses also are presented in table 9.

Table 9. Concentrations of chlorpyrifos and diazinon in water samples collected during Lagrangian and synoptic samplings of the San Joaquin River and adjacent tributaries, San Joaquin Valley, California

[µg/L, microgram per liter. <, less than. Samples collected by the California Regional Water Quality Control Board. See figure 31 for site locations.]

Site no.	Station name	Diazinon (µg/L)	Chlorpyrifos (µg/L)
	January 29–30, 1992, Lagrangian sampling		
21	Salt Slough at State Highway 165, near Stevinson	0.01	<0.01
23	Mud Slough near Gustine	<.01	<.01
24	San Joaquin River near Stevinson	.03	<.01
25	Los Banos Creek at State Highway 140	.02	<.01
26	San Joaquin River at Fremont Ford Bridge	<.01	<.01
27	Newman Wasteway near State Highway 33	.03	<.01
28	San Joaquin River near Newman	.05	.01
29	Merced River near Stevinson	.09	.02
33	Turlock Irrigation District drain lateral 5	.54	<.01
40	San Joaquin River at Laird Park	.04	<.01
43	Tuolumne River at Modesto	.04	.01
46	Ingram/Hospital Creek	.09	.01
47	San Joaquin River at Maze Road Bridge, near Modesto	.07	.01
49	Stanislaus River at Caswell State Park	.04	.01
50	San Joaquin River near Vernalis	.05	.03
	February 10–12, 1992, synoptic sampling		
31	Orestimba Creek at State Highway 33	0.26	0.02
33	Turlock Irrigation District drain lateral 5	.29	.04
39	Del Puerto Creek at State Highway 33	1.3	.03
40	San Joaquin River at Laird Park	.24	.01
46	Ingram/Hospital Creek	.24	.01
	February 17–19, 1992, synoptic sampling		
29	Merced River near Stevinson	0.32	0.05
31	Orestimba Creek at State Highway 33	.38	<.01
33	Turlock Irrigation District drain lateral 5	.5	.08
40	San Joaquin River at Laird Park	.1	.17
43	Tuolumne River at Modesto	.35	.03
49	Stanislaus River at Caswell State Park	.06	<.01
50	San Joaquin River near Vernalis	.28	.02

The results of the January sampling, when contrasted with the February samplings, indicate the effect of rainfall on increasing concentrations of organophosphate pesticides in the river system. Concentrations generally increased by more than an order of magnitude and remained high for at least 9 days following the storm. The high concentrations at most sites indicate that there must be widespread multiple sources of organophosphate pesticides to the San Joaquin River and its tributaries during the dormant-orchard spray season.

The three pesticides most frequently detected by the California Regional Water Quality Control Board were chlorpyrifos, diazinon, and parathion. Chlorpyrifos has no current drinking water standard. Although chlorpyrifos was detected frequently, the concentrations were usually less than the water-quality criterion of the U.S. Environmental Protection Agency (0.083 µg/L) for protection of aquatic life. Parathion has been banned for use in California because of concerns over its toxicity to humans during application. Some of the samples, particularly those from agricultural drains, were likely to test toxic in the three-species bioassay test of the U.S. Environmental Protection Agency. Those samples had concentrations of diazinon, which is toxic to *Ceriodaphnia dubia,* one of the test organisms, in excess of the published value of 0.35 µg/L (Amato and others, 1992). However, median concentrations of diazinon generally are less than that value. Most concentrations at integrator sites were less than 0.35 µg/L with the exception of a few collected during the February 1992 synoptic sampling (table 9).

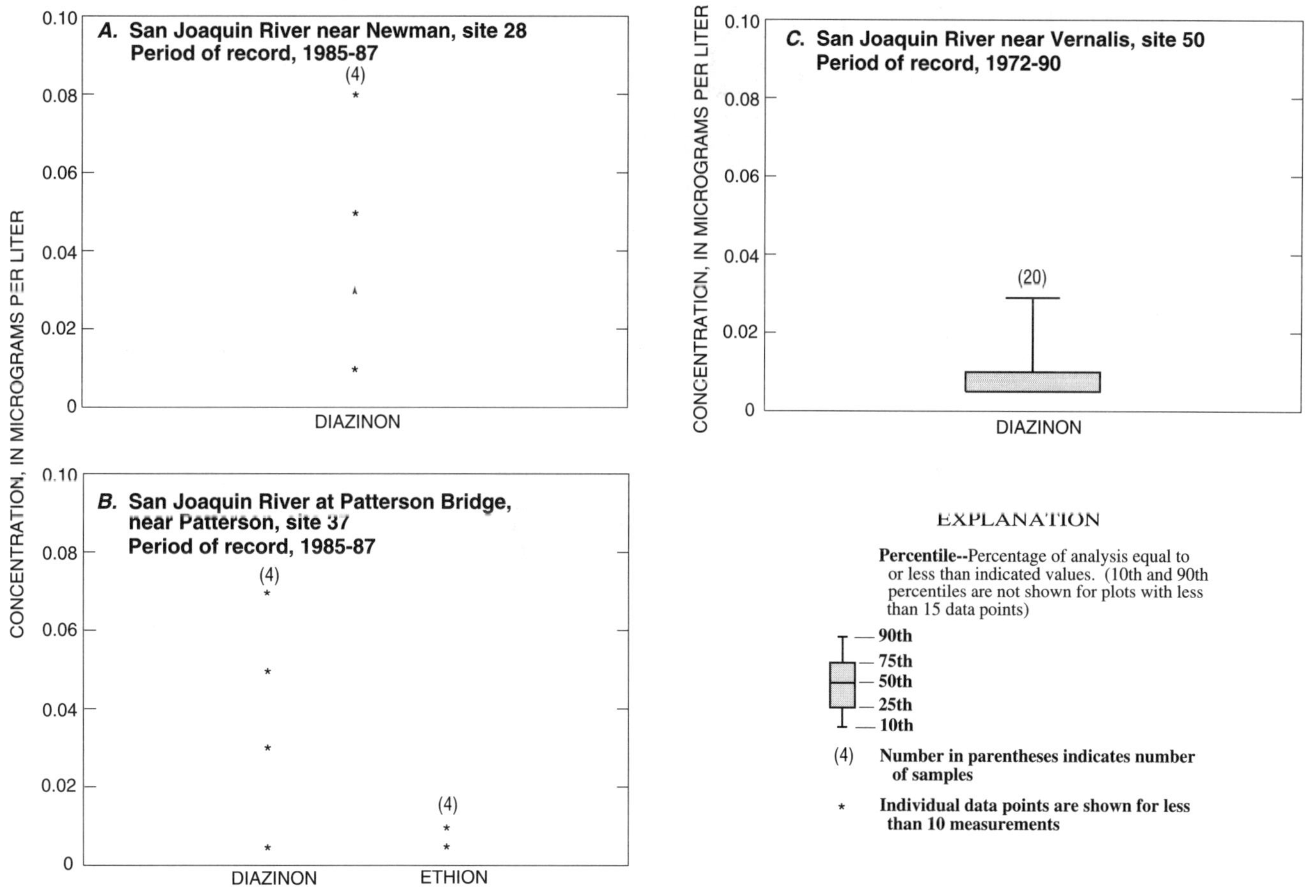

Figure 23. Concentrations of organophosphate pesticides in water samples collected from the San Joaquin River near Newman, near Patterson, and near Vernalis, San Joaquin Basin, California. (Compiled from U.S. Geological Survey data base. See figure 21 for site locations.)

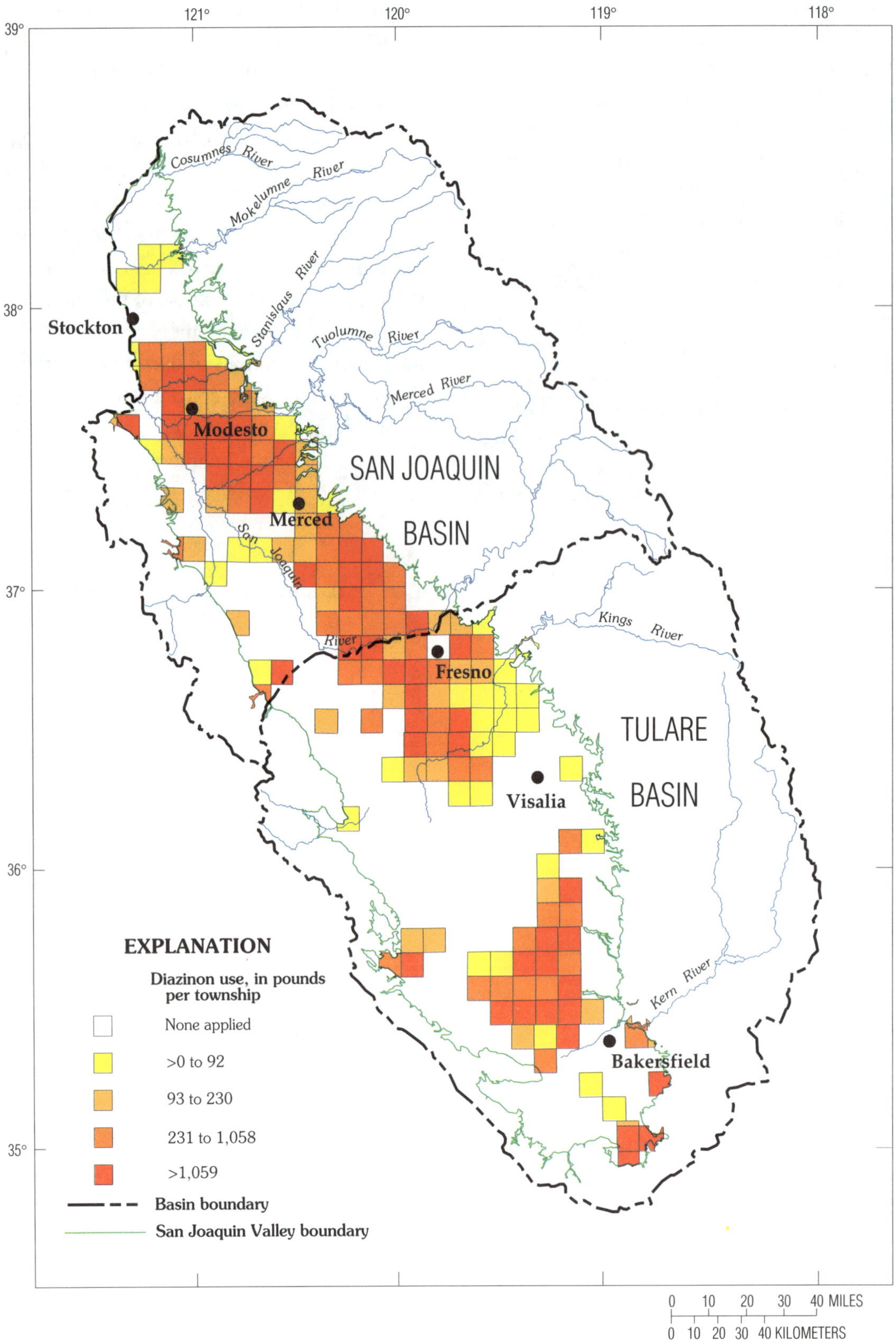

Figure 24. Amount of diazinon applied to almond orchards in the San Joaquin Valley, California, 1990. (Data from California Environmental Protection Agency, Department of Pesticide Regulation well-inventory data base, Maes and others, 1991.)

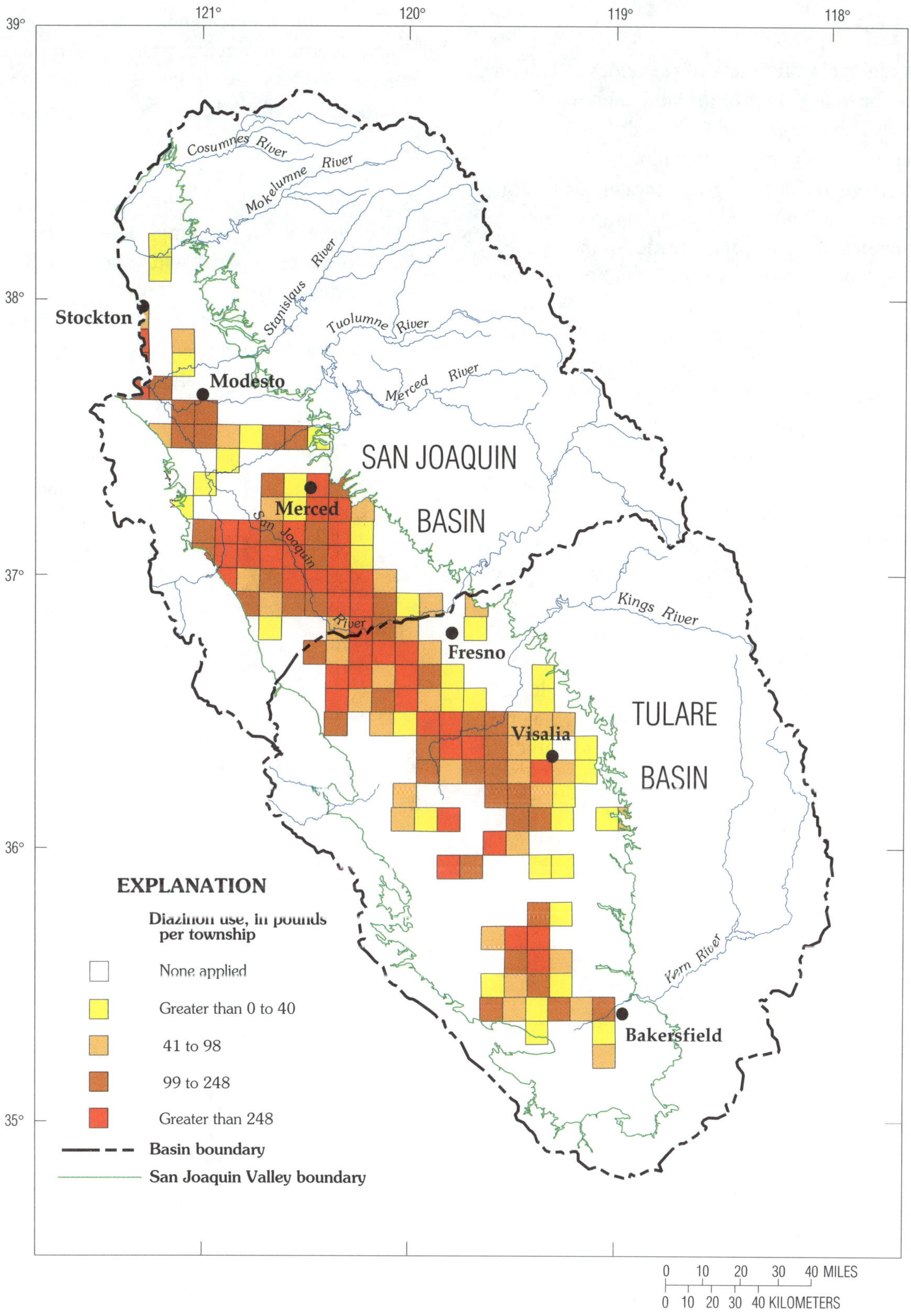

Figure 25. Amount of diazinon applied to alfalfa fields in the San Joaquin Valley, 1990. (Data from California Environmental Protection Agency, Department of Pesticide Regulation well-inventory data base, Maes and others, 1991.)

Ground Water

A regional-scale study of pesticides in ground water for the valley floor of the study unit was reported by Domagalski and Dubrovsky (1991, 1992). This section is a synopsis of that report.

Pesticide residues in ground water can be attributed largely to soil properties, physiographic provinces, chemical or physical properties of the pesticides, types of pesticides used, land use or cropping pattern, and depth to ground water. Most ground-water pesticide residues were detected on the east side of the valley. These residues were attributed to sandy or coarse-grained soils of Sierra Nevada provenance, a relatively shallow ground-water table (less than 100 ft below land surface) in some subareas, and the use of water-soluble pesticides with long environmental half-lives. In contrast, soils of the western valley and within the basin areas are fine grained and are of Coast Ranges provenance. Despite a shallow ground-water table in much of that area (within 20 ft of land surface), there were relatively few detections. The lack of detections was attributed to a much longer residence time of pesticides in the fine-grained sediments of the unsaturated zone and slow velocity of water recharge. The longer residence time allows sufficient time for degradation reactions to take place.

The detected pesticides include atrazine, bromacil, 2,4-DP, diazinon, dibromochloropropane (DBCP), 1,2-dibromoethane, dicamba, 1,2-dichloropropane, diuron, prometon, prometryn, propazine, and simazine. With the exception of diazinon, all these pesticides are applied directly to the soil and not to the vegetation.

The most frequently detected pesticide was DBCP, which was detected in water samples from 1,419 of the 4,507 wells sampled during 1971-88. The areal distribution of wells with detected DBCP and with no detected DBCP are shown in figure 41. Those samples were collected by various State agencies, primarily the California Department of Health Services. Data were retrieved from the well-inventory data base

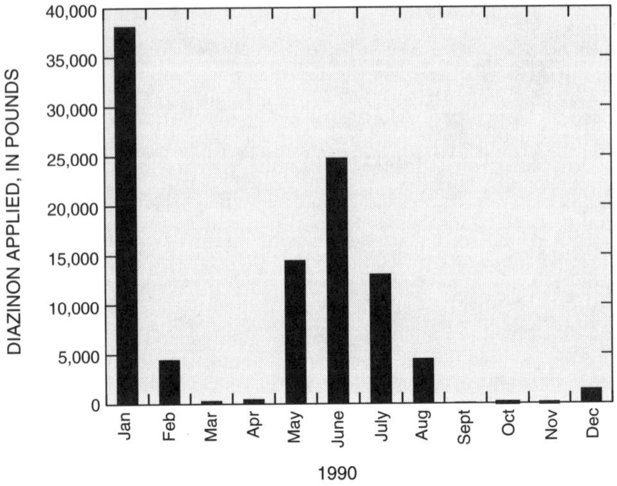

Figure 26. Monthly application of diazinon on almond orchards in the San Joaquin Valley, California, 1990. (Data from California Environmental Protection Agency, Department of Pesticide Regulation.)

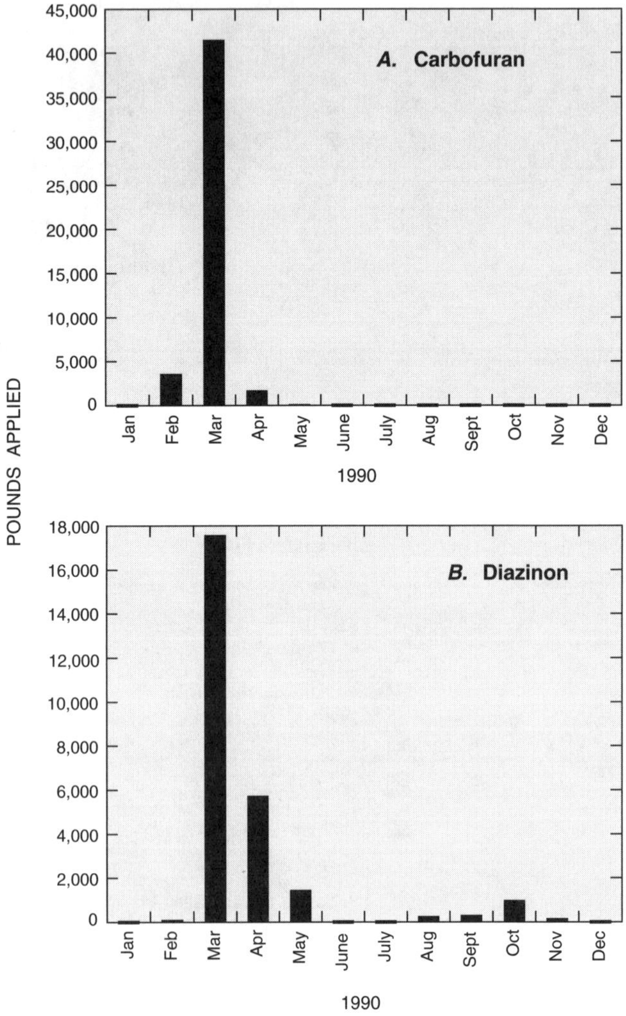

Figure 27. Monthly application of carbofuran and diazinon on alfalfa fields in the San Joaquin Valley, California, 1990. (Data from California Environmental Protection Agency, Department of Pesticide Regulation well-inventory data base, Maes and others, 1991.)

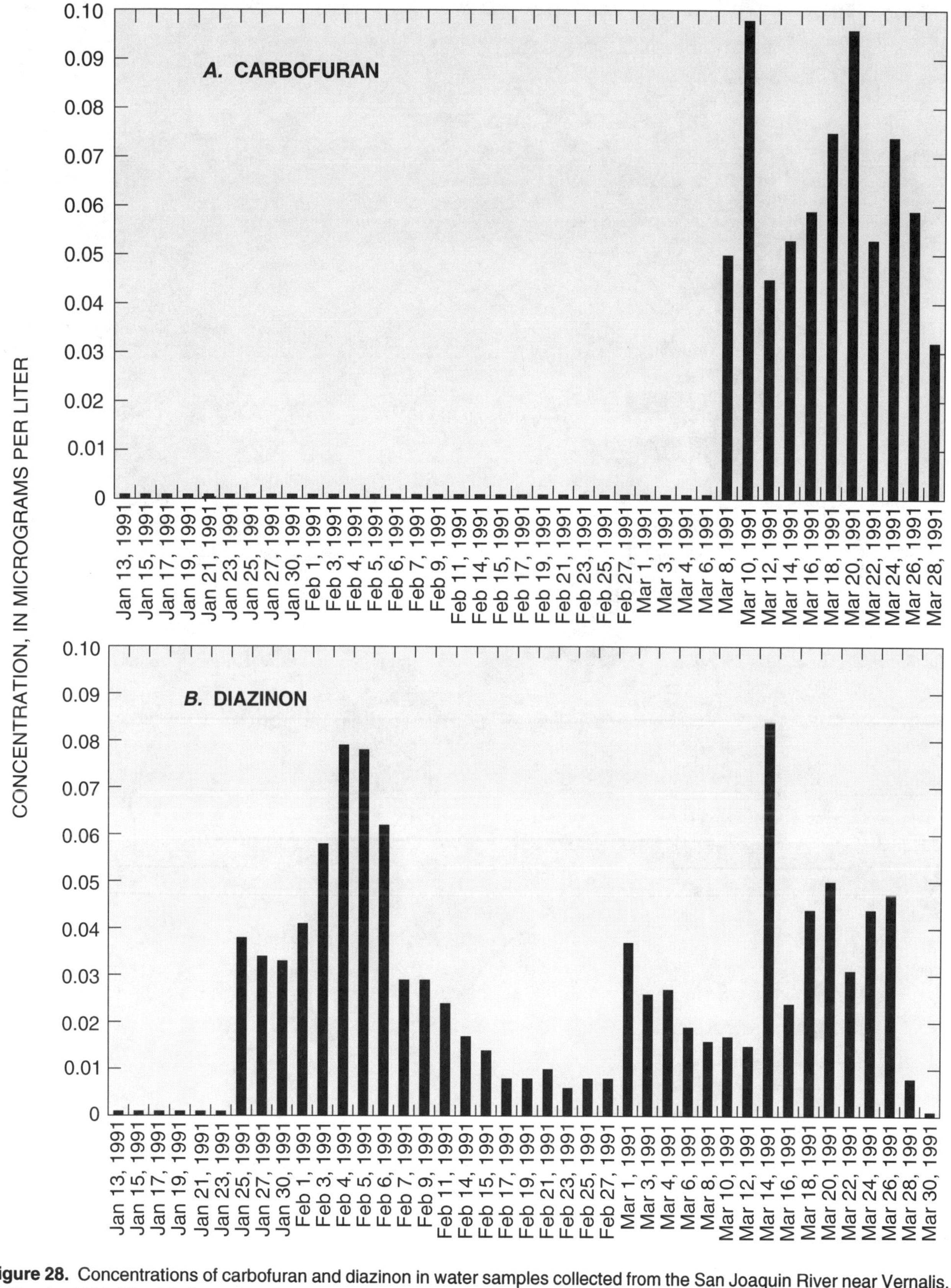

Figure 28. Concentrations of carbofuran and diazinon in water samples collected from the San Joaquin River near Vernalis, California, 1991 (Site 50, figure 21. Crepeau and others, 1991.)

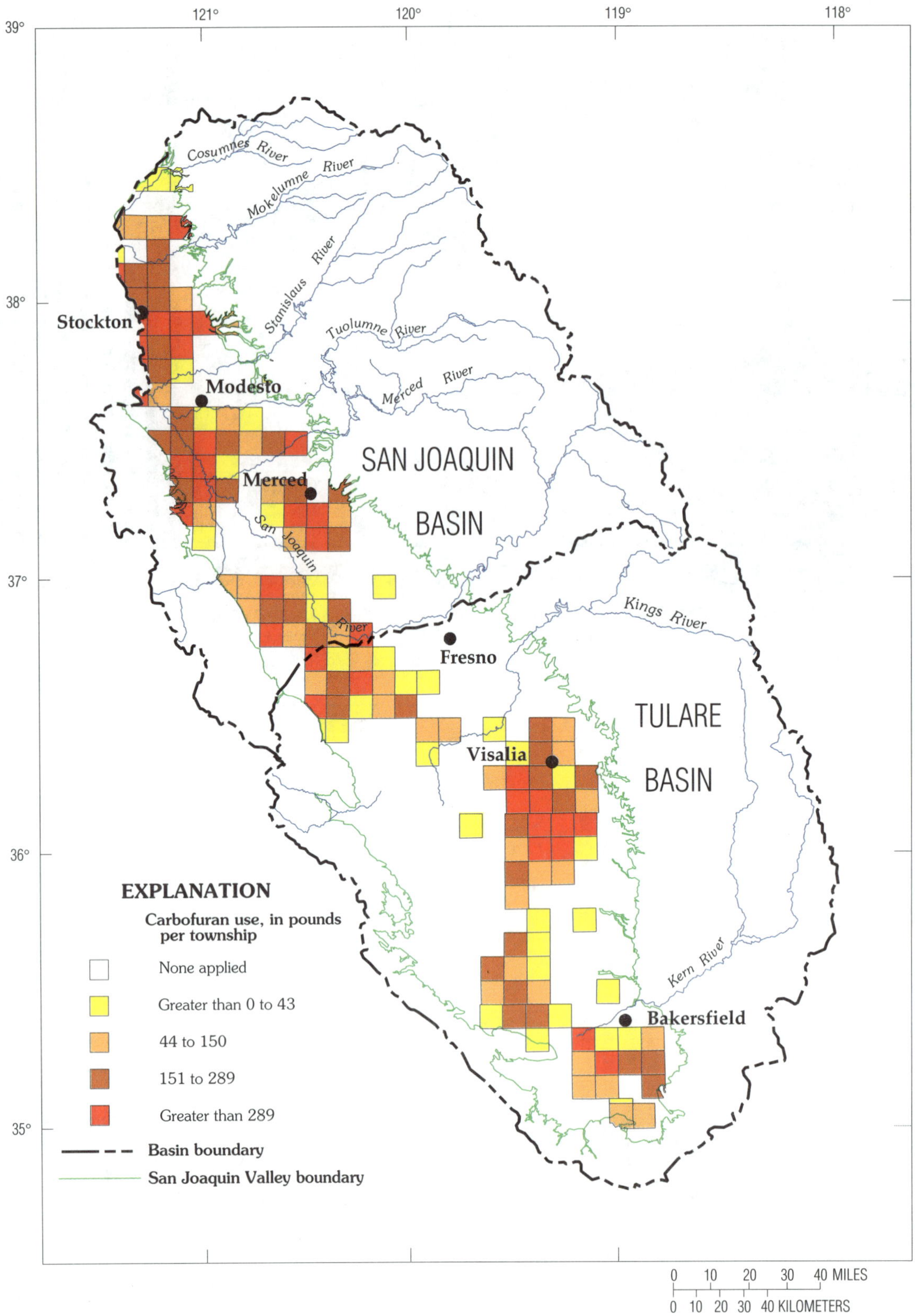

Figure 29. Amount of carbofuran applied to alfalfa fields in the San Joaquin Valley, Caifornia, 1990. (Data from California Environmental Protection Agency, Department of Pesticide Regulation well-inventory data base, Maes and others, 1991.)

of the California Department of Pesticide Regulation (Maes and others, 1991). DBCP concentrations are shown in figure 42. The median concentration is 2.0 µg/L. The California Department of Health Services has established a maximum contaminant level of 0.2 µg/L for DBCP in drinking water.

Land use for urban areas, cropland and pasture, orchards and vineyards, and wetlands in the San Joaquin Valley is shown in figure 43. These land-use categories are based on 1975 land-use data from the geographical information retrieval and analysis system (GIRAS) data base of the USGS (1986). DBCP, a soil fumigant used to control nematodes, was used primarily in orchards and vineyards, but also was used on row crops. Most DBCP detections were in the orchard-vineyard land-use category. DBCP has essen-

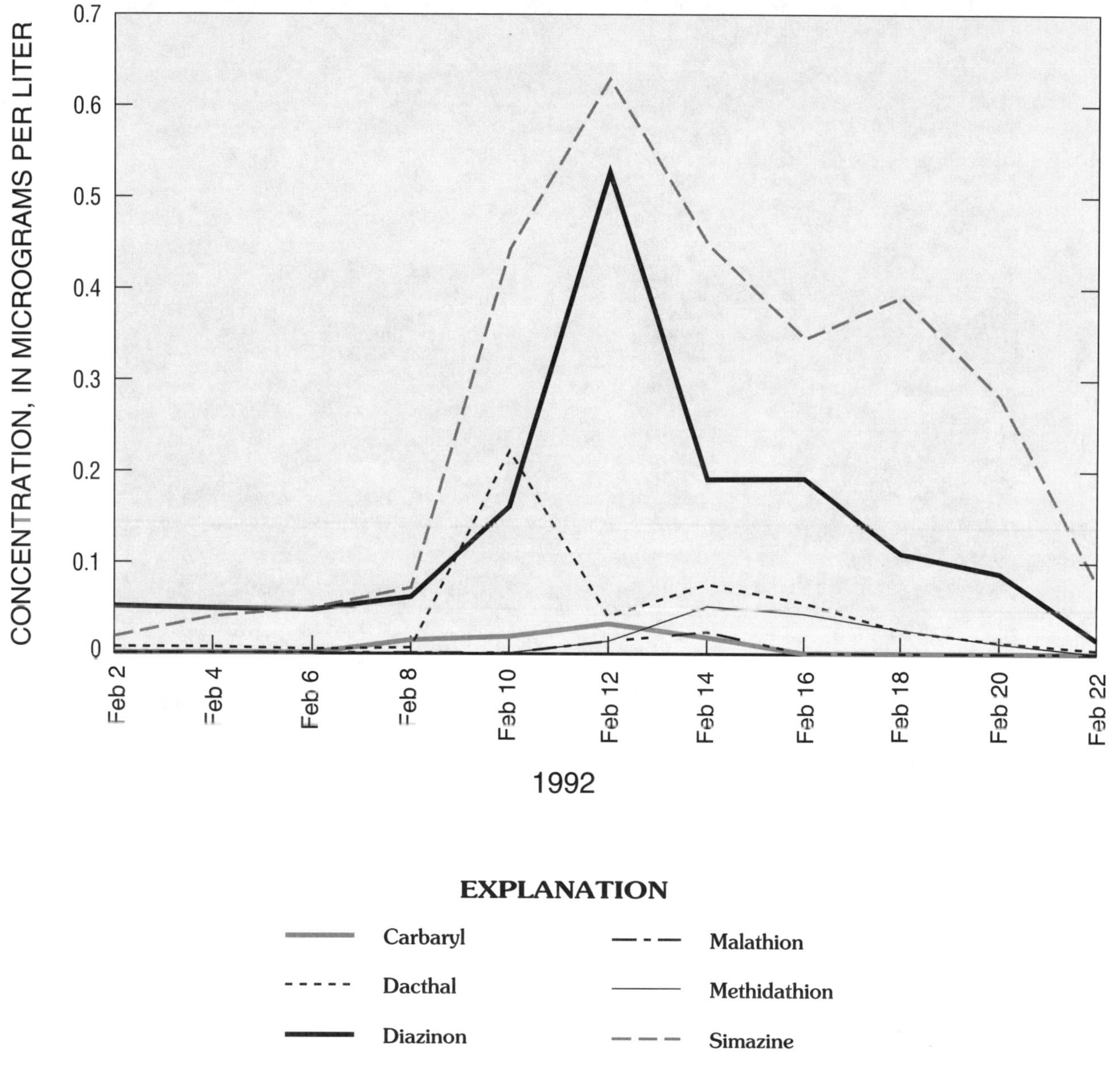

Figure 30. Concentrations of selected pesticides in water samples collected from the San Joaquin River near Vernalis, California, during a February 1992 storm. (Site 50, figure 31. Compiled from U.S. Geological Survey data base.)

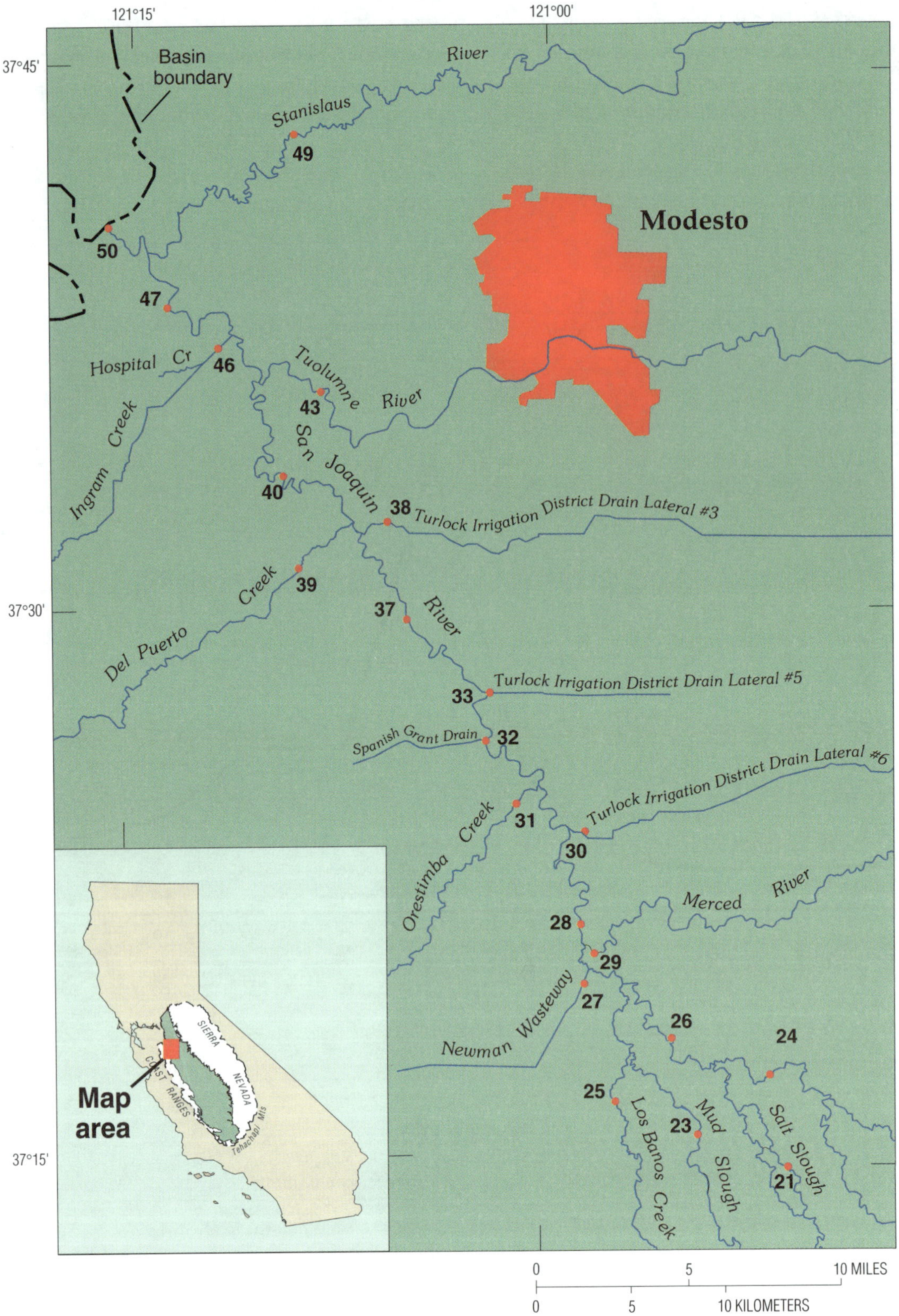

Figure 31. Locations of sites sampled by the Caifornia Regional Water Quality Control Board, 1991-92—San Joaquin River, California, and adjacent agricultural drains and tributaries. (See table 6 for station numbers and descriptions.)

tially no phytotoxicity, but was banned by the California Department of Food and Agriculture in 1977 because of its occurrence in ground water and its toxicity to humans.

DBCP was applied directly to soils in California between 1955 and 1977. Because of its volatility, the fields were either covered or irrigated. DBCP is soluble in water (solubility is 1.2 g/L). The most significant property of DBCP in ground water is its resistance to degradation in water; the best current estimate of a half-life is 6.1 years (Deeley and others, 1991). The proximity of DBCP detections in ground water near the city of Fresno is a concern, as the ground water is used for drinking water.

Other soil fumigants have been detected in ground water of the study area. In 709 wells sampled, there were 29 detections of 1,2-dibromoethane and in 900 wells sampled, 19 detections of 1,2-dichloropropane (Domagalski and Dubrovsky, 1991, 1992). Both compounds have been banned in California. No other fumigant has replaced DBCP. The fumigants currently being used, such as methyl bromide and 1,3-dichloropropene, are applied to new orchards or vineyards prior to planting. Methyl bromide (763 wells) and 1,3-dichloropropene (823 wells) were not detected.

Organonitrogen herbicides, such as bromacil and diuron, were detected in well water in Tulare

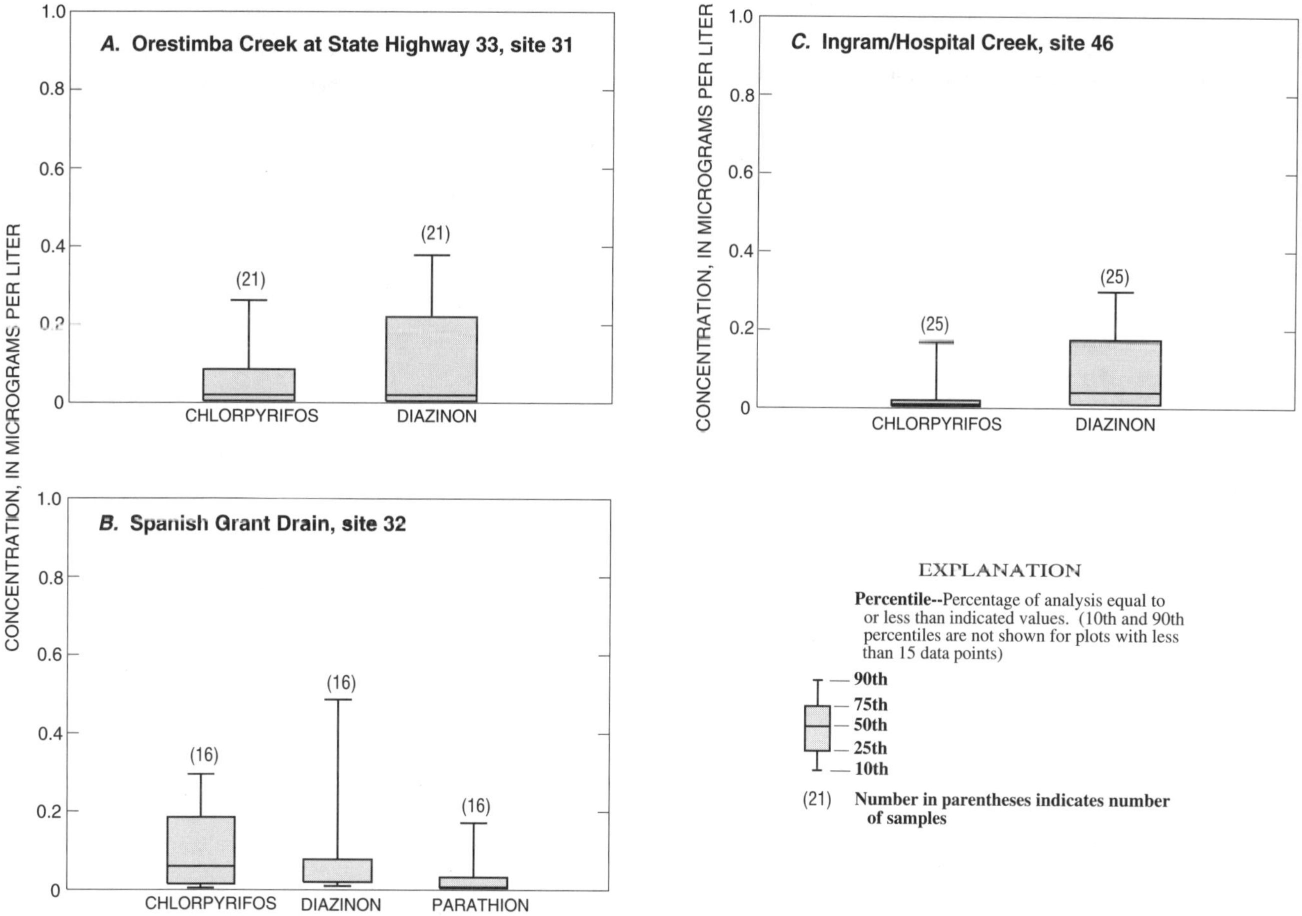

Figure 32. Concentrations of selected organophosphate pesticides in water samples collected from the San Joaquin River and adjacent agricultural drains and tributaries, San Joaquin Basin, California: Valley, west side, alluvial deposits environmental setting—agriculture (row crops). (Samples collected by the California Regional Water Quality Control Board, 1991-92. See figure 31 for site locations.)

County (Troiano and Segawa, 1987) and other locations in the San Joaquin Valley. These detections are reported in the California well-inventory data base. Bromacil is used primarily in orange orchards (figs. 44 and 45), and most detections were in the easternmost part of the San Joaquin Valley within Tulare County orange orchards (fig. 45*B*).

Selected wells were sampled along two transects in Tulare County, which included the major orange orchard area. Samples were analyzed for carbamate pesticides, organonitrogen herbicides, organophosphate pesticides, and triazine herbicides. Atrazine, bromacil, diuron, and simazine were detected. The median concentration of bromacil in wells of the San Joaquin Valley with detectable amounts is 1.9 µg/L (fig. 46). There is no current drinking-water standard for bromacil.

In contrast to bromacil, diuron is used extensively throughout the San Joaquin Valley. Diuron

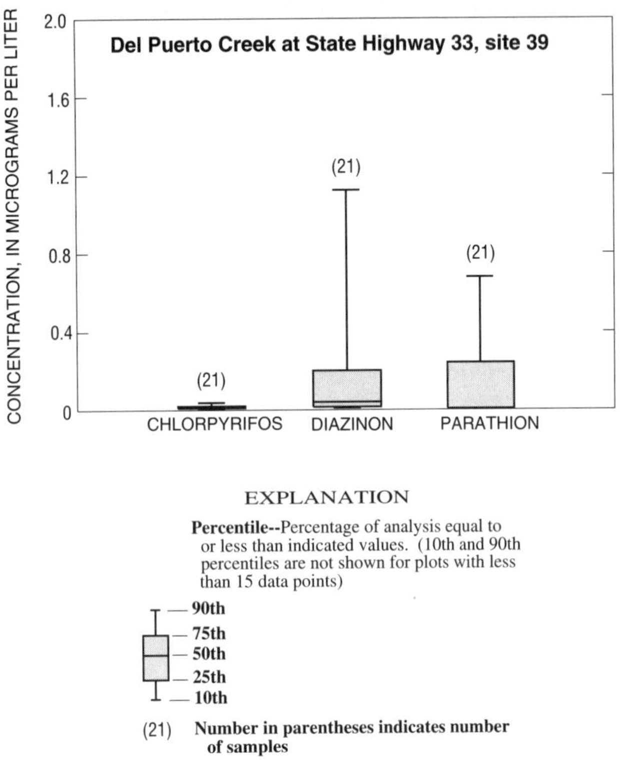

Figure 33. Concentrations of selected organophosphate pesticides in water samples collected from the San Joaquin River and adjacent agricultural drains and tributaries, San Joaquin Basin, California: Valley, west side, alluvial deposits environmental setting—agriculture (orchards). (Samples collected by the California Regional Water Quality Control Board, 1991-92. See figure 31 for site locations.)

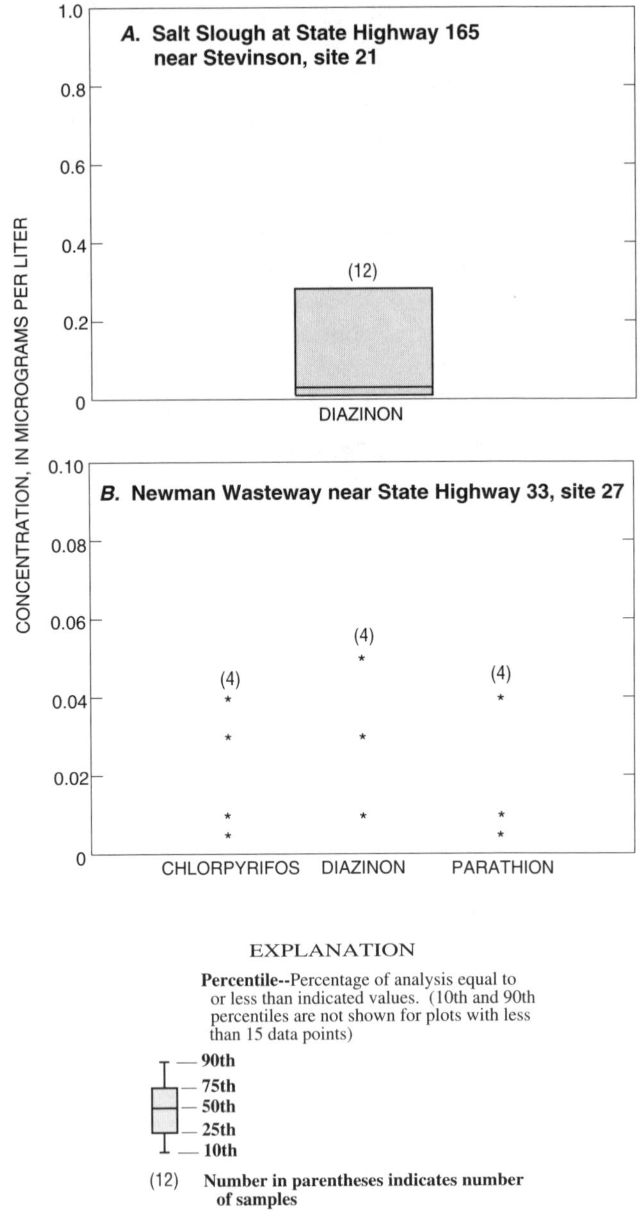

Figure 34. Concentrations of selected organophosphate pesticides in water samples collected from the San Joaquin River and adjacent agricultural drains and tributaries, San Joaquin Basin, California: Valley, west side, alluvial deposits environmental setting—agriculture (field crops) . (Samples collected by the California Regional Water Quality Control Board, 1991-92. See figure 31 for site locations.)

also is used on orange orchards, and monitoring has been more extensive than for bromacil. Diuron was detected only in water from wells near the orchards (figs. 47 and 48). However, diuron also is used for weed control on roadways, and the possibility that these applications also contribute to leaching to the ground water cannot, at present, be ignored. The median concentration of diuron detected in wells was 0.62 µg/L (fig. 49). There is no current drinking-water standard for diuron.

The USGS has sampled the shallow ground water of the western San Joaquin Valley and has completed a regional study of pesticides in the regional aquifer of the San Joaquin Valley (Domagalski and Dubrovsky, 1991, 1992). The objectives of the reconnaissance of pesticide residues in shallow ground water of the western San Joaquin Valley were to determine (1) the extent of pesticides in shallow ground water and subsurface agricultural drainwater, (2) the distribution of pesticides in shallow ground water, and (3) the implications for assessing pesticides in ground water on a regional scale. Because of a long history of

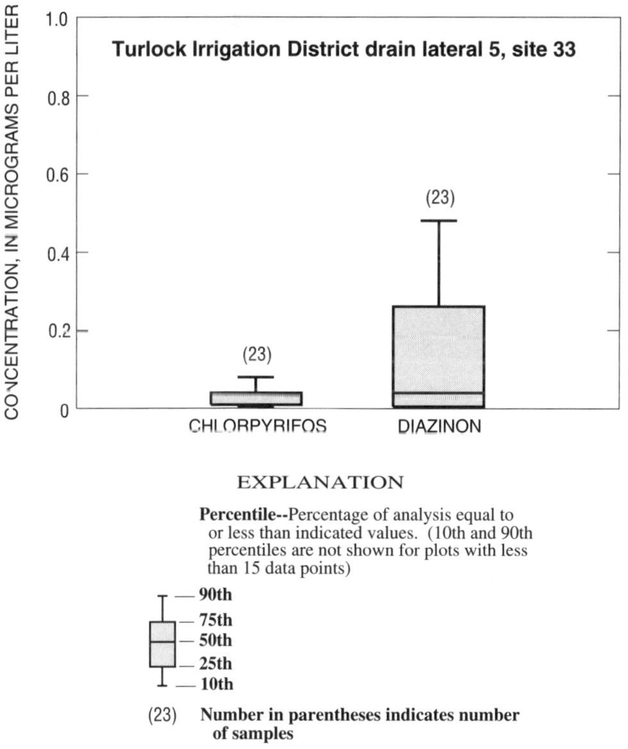

Figure 35. Concentrations of selected organo- phosphate pesticides in water samples collected from the San Joaquin River and adjacent agricultural drains and tributaries, San Joaquin Basin, California: Valley, east side, alluvial deposits environmental setting—agriculture (orchards) . (Samples collected by the California Regional Water Quality Control Board, 1991-92. See figure 31 for site locations.)

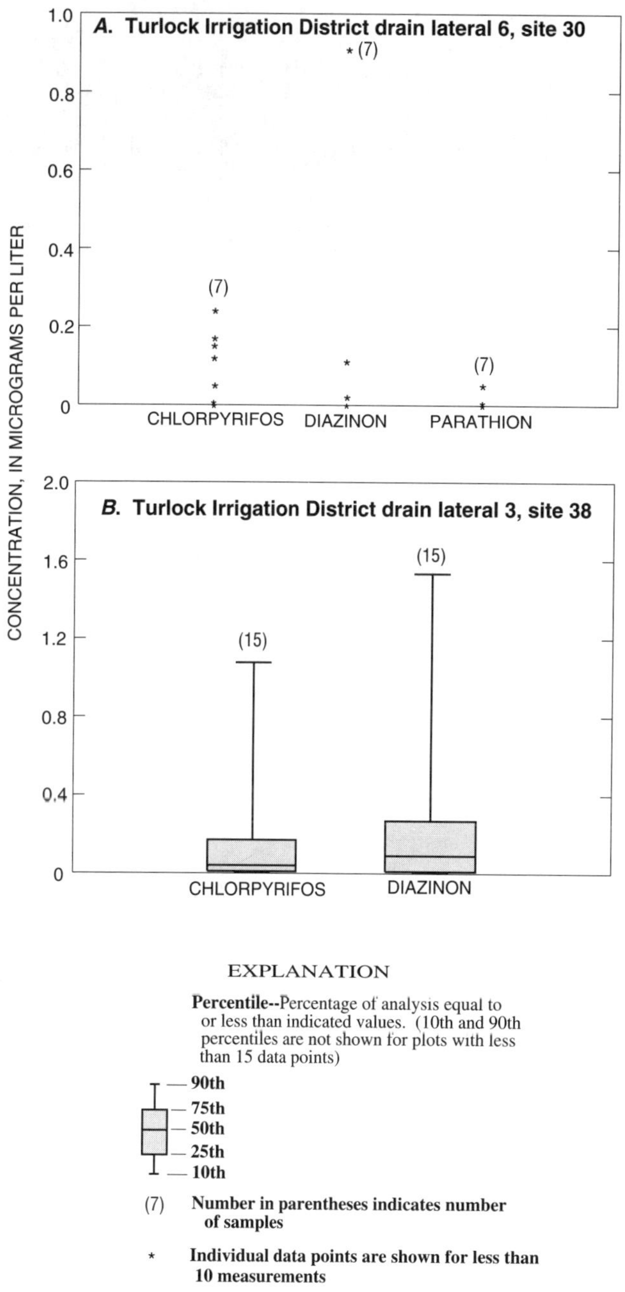

Figure 36. Concentrations of selected organo- phosphate pesticides in water samples collected from the San Joaquin River and adjacent agricultural drains and tributaries, San Joaquin Basin, California: Valley, east side, alluvial deposits environmental setting—agriculture (field crops). (Samples collected by the California Regional Water Quality Control Board, 1991-92. See figure 31 for site locations.)

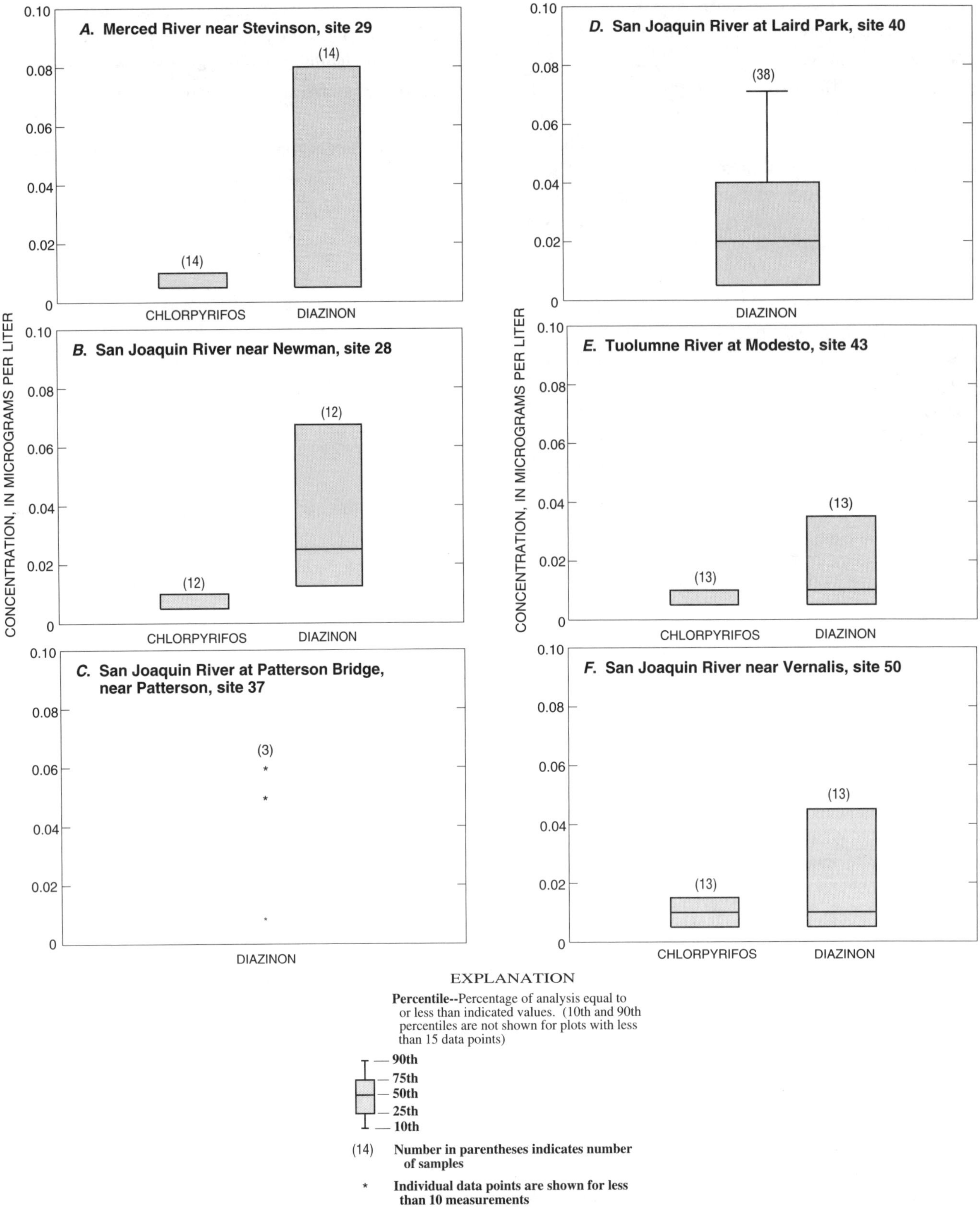

Figure 37. Concentrations of selected organophosphate pesticides in water samples collected at San Joaquin Basin, California, integrator sites. (Samples collected by the California Regional Water Quality Control Board, 1991-92. See figure 31 for site locations.)

irrigation, ground water occurs within a few feet of land surface in much of the western San Joaquin Valley and parts of the Tulare Basin, within soil and alluvial sediments (Deverel and others, 1984). Salinity is high in shallow ground water because of evapotranspiration. Subsurface tile drains have been installed to remove the saline water because of its adverse effect on crop productivity.

In 1984, 80 observation wells (screened at 20 ft below land surface), 32 field drains, and 10 collector drains that collect water from several field drains were sampled; 24 wells and 14 drains were resampled in 1985. These sites were primarily in the distal reaches of alluvial fans of the central part of the western valley. In 1990, five observation wells in the southern part of the valley were sampled. For the 1984 study, the detection limit for the analysis of triazine herbicides and organophosphate pesticides was 1 μg/L. The 1985 study included specific analyses for triazine herbicides, with a detection limit of 0.1 μg/L. Triazine herbicides were analyzed in the 1990 study, with a detection limit of 0.05 μg/L (Domagalski and Dubrovsky, 1991, 1992).

Triazine herbicides and organophosphate pesticides were not detected in subsurface drains, which also collect water from layers below the water table (Deverel and others, 1989). Therefore, relative to the water table, older water was sampled, and any pesticides present probably were degraded.

Prometon and simazine were detected in water samples from three observation wells in the central part of the western valley in 1984. Prometon and simazine also were detected in 9 of 24 observation wells in 1985, and atrazine was detected in 1 well. These three triazine herbicides are used to control weeds along roads, but are not used on crops in this area. All the observation wells were at the edges of agricultural fields near roads and, therefore, near roadside spraying. Cotton is one of the major crops in this part of the study unit, but neither of the triazine herbi-

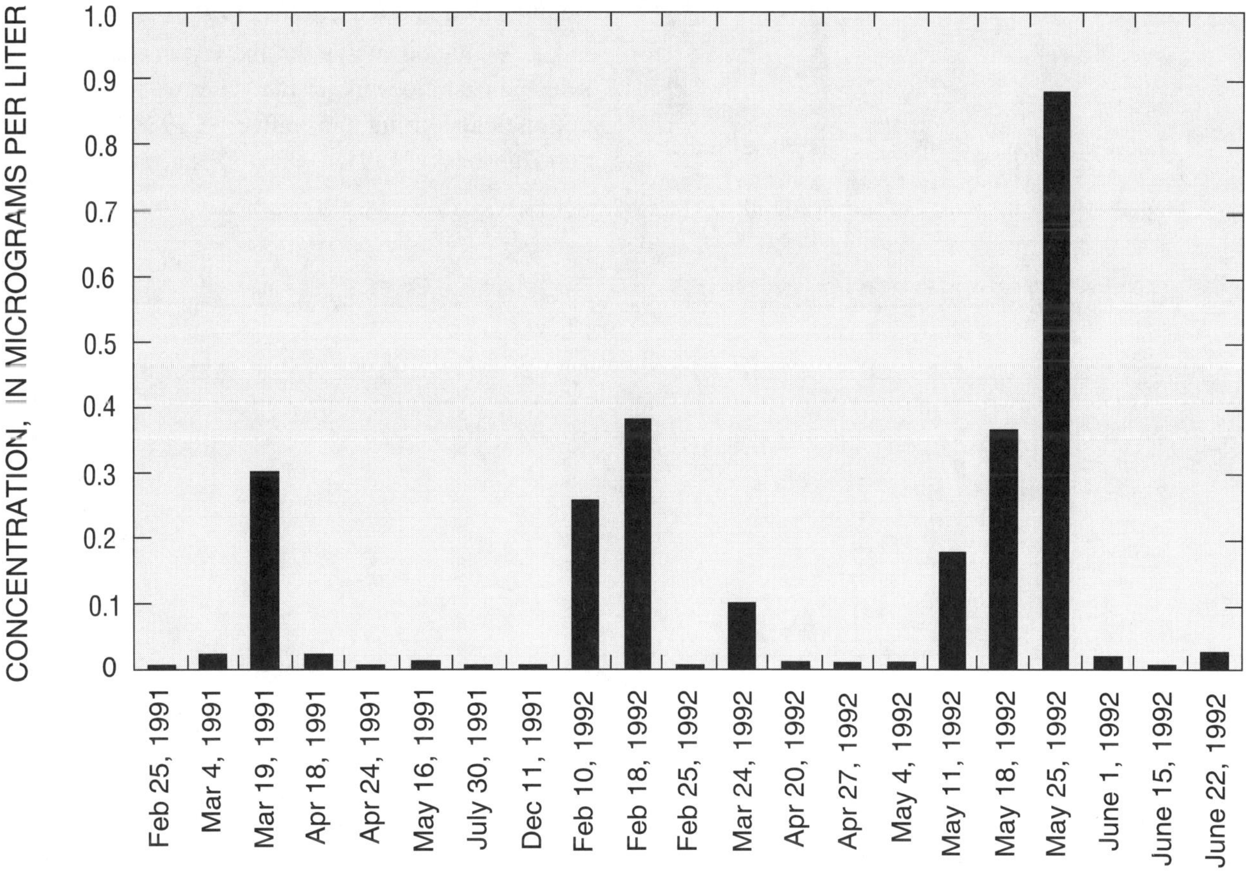

Figure 38. Concentrations of diazinon in water samples collected at the Orestimba Creek at California State Highway 33 integrator site, San Joaquin Basin, California. (Site 31, figure 31. Samples collected by the California Regional Water Quality Control Board, 1991-92.)

cides used on cotton—cyanazine and prometryn—were detected in ground water in this area.

Atrazine, diazinon, prometon, prometryn, and propazine were detected in water samples from four of the five shallow observation wells in the southern part of the valley. Prometryn is a pre-emergent herbicide used on cotton fields. Diazinon, an organophosphate pesticide, is used frequently throughout the valley on various crops. The remaining pesticides—atrazine, prometon, and propazine—are triazine herbicides used to control weeds along roads in this area.

Although limited in scope, the reconnaissance of shallow observation wells in the southern part of the valley indicates that some pesticides used on cotton, most notably prometryn, have leached to the shallow water table. However, only limited areas are susceptible because of the widespread clay soils derived from lacustrine sediments. Reconnaissance studies of the central-western and southern parts of the valley indicate that triazine herbicides used for weed control on roads have leached to the shallow ground water (Domagalski and Dubrovsky, 1991, 1992).

A detailed analysis of the vertical distribution of selected pesticides in ground water was done in two cotton fields during the summer of 1988 (Domagalski and Dubrovsky, 1991, 1992). The primary purpose of

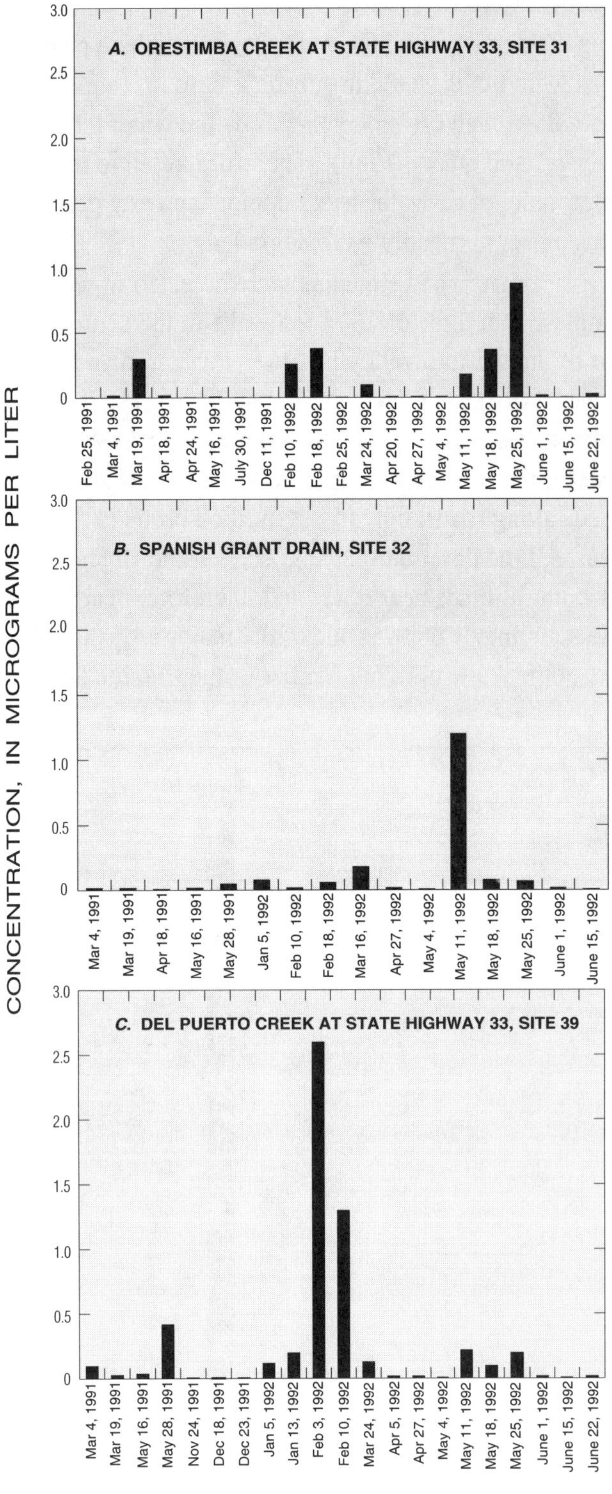

Figure 39. Concentrations of diazinon in water samples collected at Orestimba Creek at California State Highway 33, Spanish Grant Drain, and Del Puerto Creek at State Highway 33, San Joaquin Basin, California. (Samples collected by the California Regional Water Quality Control Board, 1991-92. See figure 31 for site locations.)

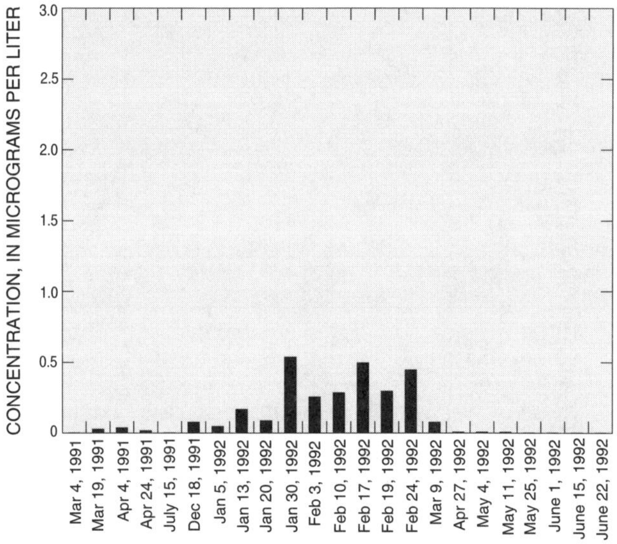

Figure 40. Concentrations of diazinon in water samples collected at the Turlock, California, Irrigation District drain lateral 5 integrator site. (Site 33, figure 31. Samples collected by the California Regional Water Quality Control Board, 1991-92.)

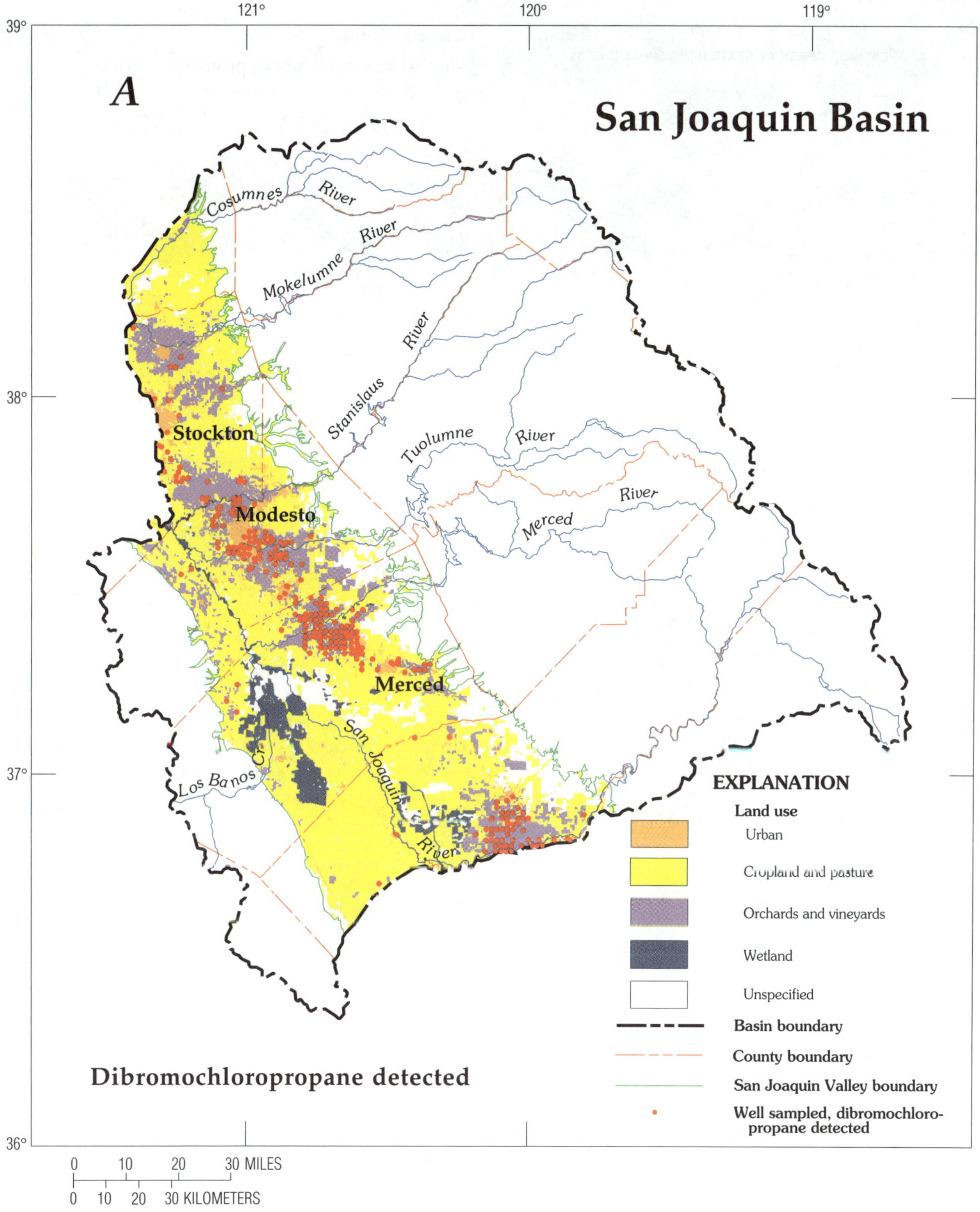

Figure 41. Areal distribution of wells sampled for dibromochloropropane, 1971-88, and partial land use, 1975, San Joaquin Valley, California. (Data from California Environmental Protection Agency, Department of Pesticide Regulation well inventory data base, Maes and others, 1991; land use data from U.S. Geological Survey, Geographical Information Retrieval and Analysis System data base, 1986.)

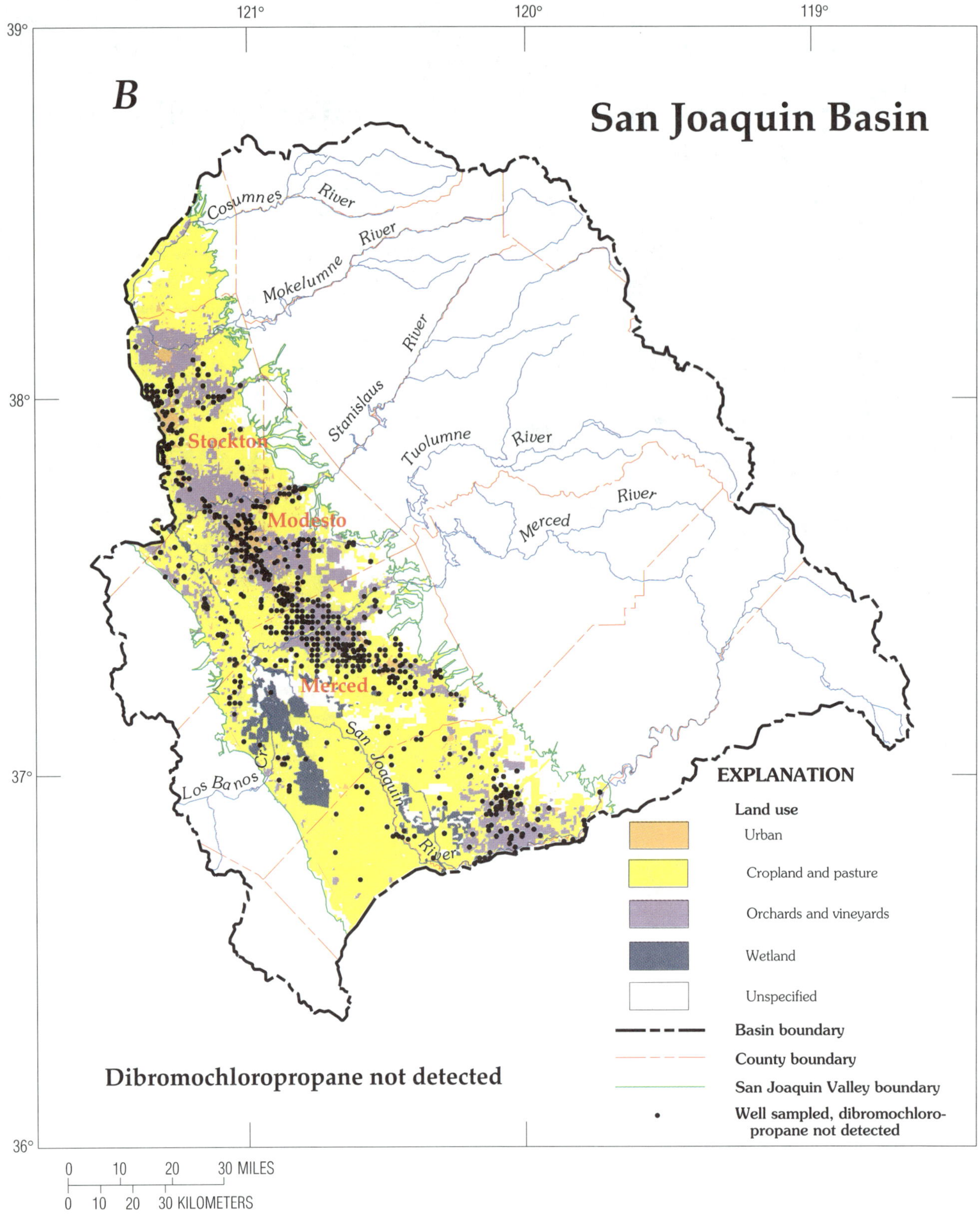

Figure 41—Continued.

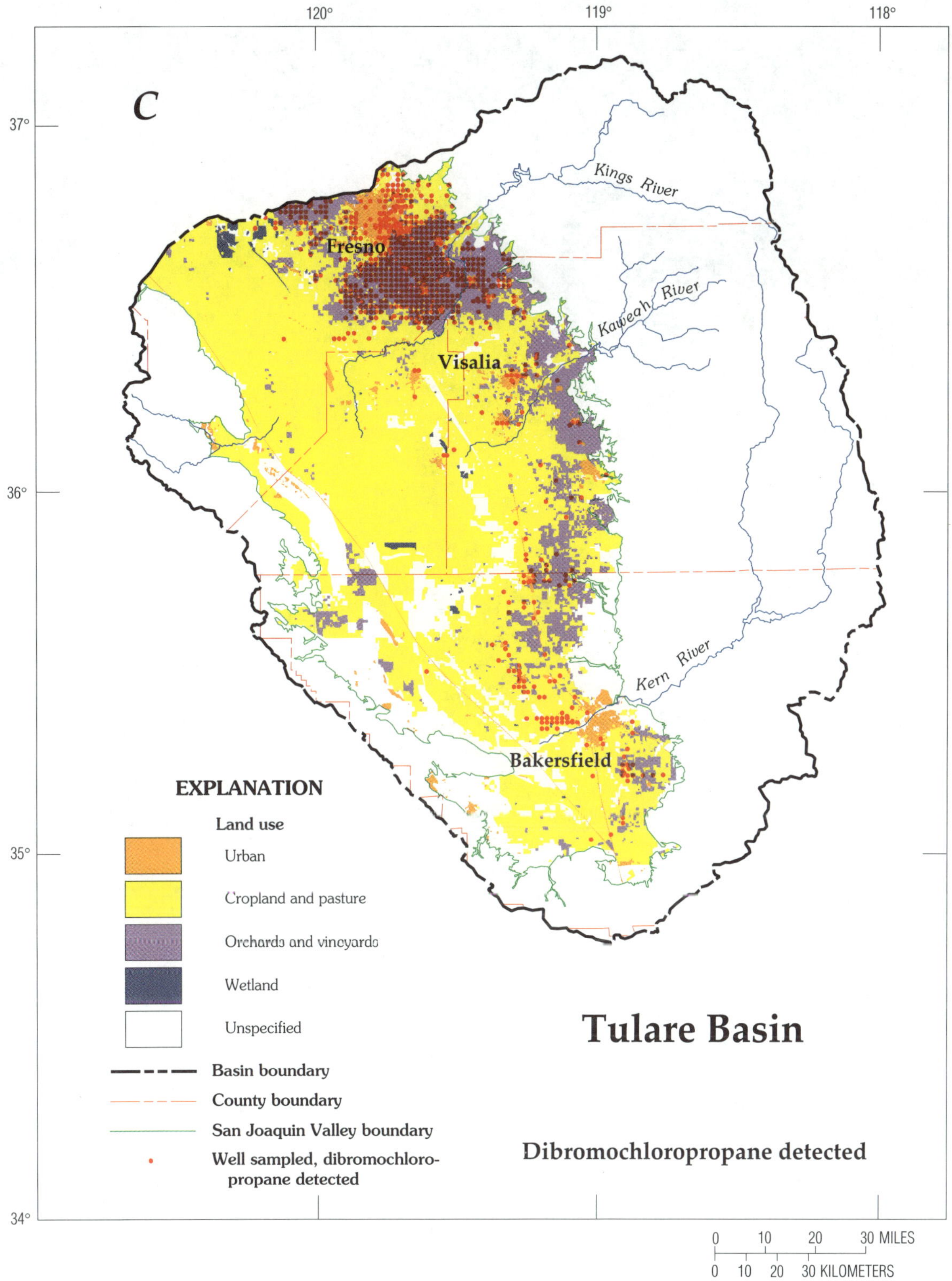

Figure 41—Continued.

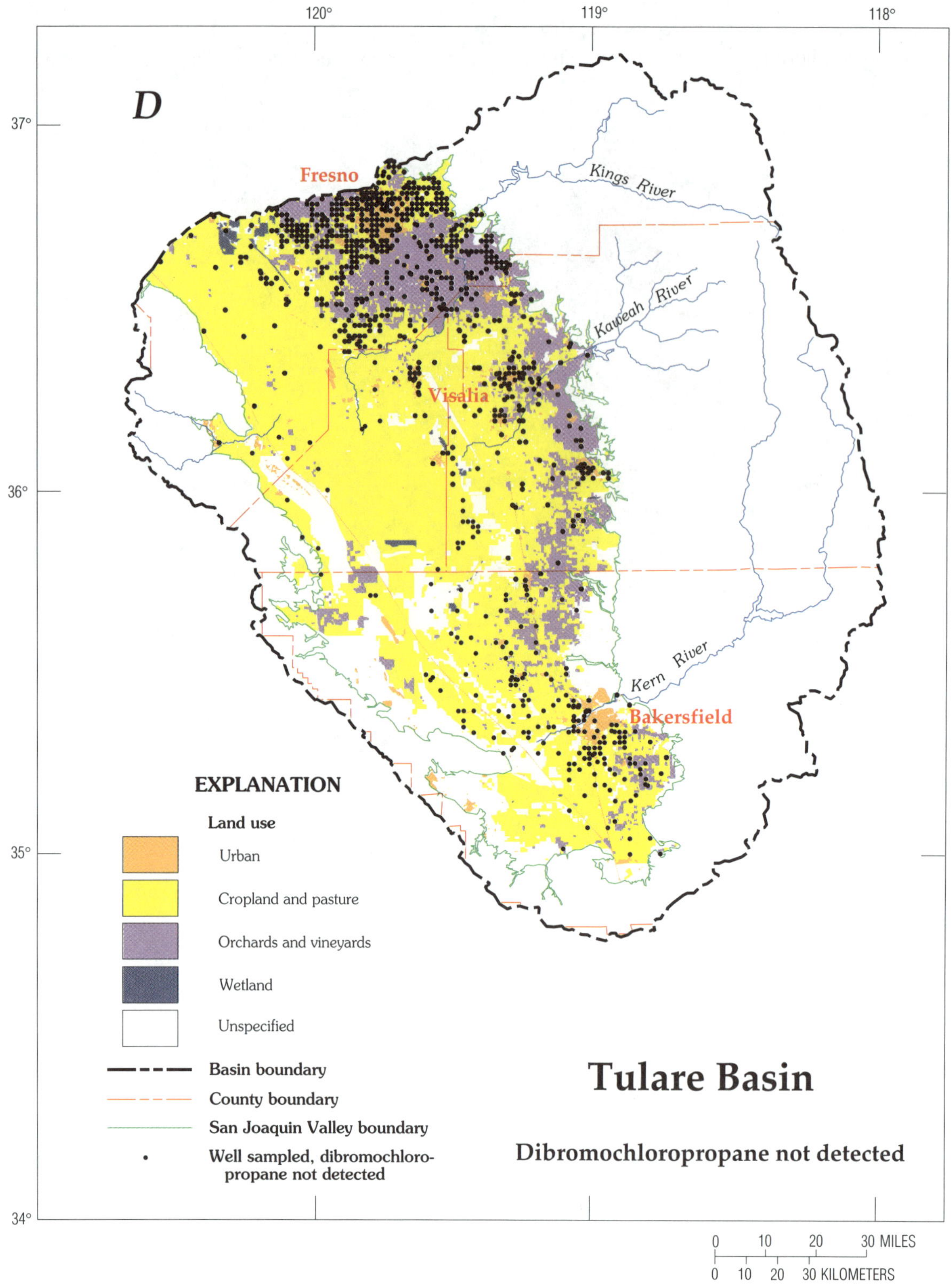

Figure 41—Continued.

this analysis was to estimate the leaching potential of pesticides associated with cotton, a major crop of the central-western and southern parts of the valley. The two cotton fields are in western Fresno County in the lower parts of an extensive alluvial fan of Coast Ranges origin. Thirty-seven wells were sampled with screens at discrete intervals: 9 wells were 10 ft deep, 10 wells were 20 ft deep, 8 wells were 30 ft deep, 4 wells were 40 ft deep, and 6 wells were 50 ft deep. The wells were installed in the cotton fields, presumably away from the influence of roadside spraying. Samples were analyzed for triazine herbicides and organophosphate and carbamate pesticides. The detection limits were 0.1, 0.01, and 0.2 µg/L, respectively (Wershaw and others, 1987). Results indicated that pesticides used on cotton were not leaching to the ground water of these two fields. The results are attributed to possible sorption of pesticides on the abundant clay-sized particles of the near-surface alluvial sediments.

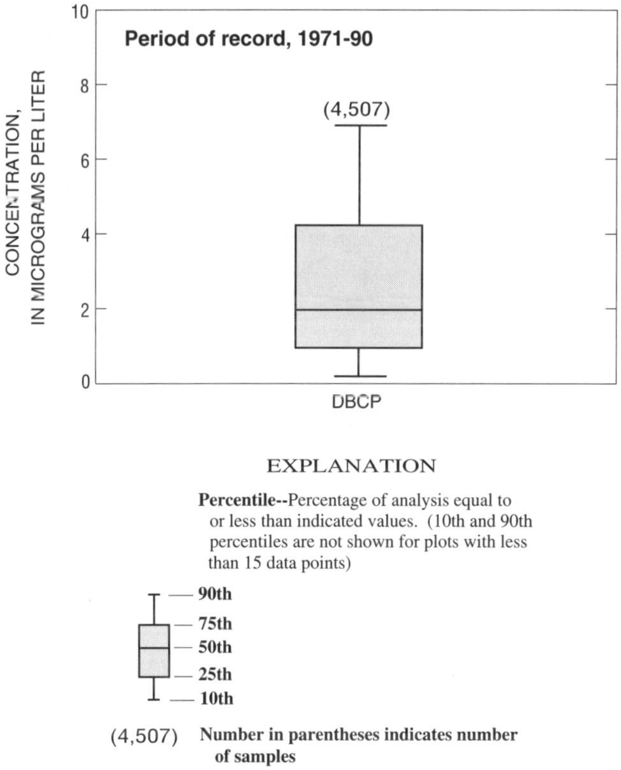

Figure 42. Concentrations of dibromochloropropane in ground water of the San Joaquin Valley, California. (Data from California Environmental Protection Agency, Department of Pesticide Regulation well-inventory data base, Maes and others, 1991. See figure 41 for site locations.)

The reconnaissance of San Joaquin Valley ground water was designed to sample the upper and lower zones of the water-supply aquifer in the western and southern parts of the valley and to sample wells screened near the water table in the east side of the valley where no confining unit is present. The upper and lower zones of the western and southern parts of the valley are separated by the Corcoran Clay Member of the Tulare Formation (fig. 3). Ground water is present under partly confined conditions in the upper zone. In addition to pesticides, selected wells were sampled for tritium to assess the relative age of the ground water sampled. The tritium content of meteoric water increased after 1952 because of atmospheric testing of nuclear weapons (Davis and Bentley, 1982). In the San Joaquin Valley, ground water with high tritium content is interpreted as recent (post-1952) irrigation recharge.

The areal distribution of wells sampled for pesticide residues in the San Joaquin Valley ground water study is shown in figure 50, and well depths and uses of the well water are given by Domagalski and Dubrovsky (1991, 1992). Wells of the west side of the valley were sampled in 1985, the south side of the valley in 1986, and the east side of the valley in 1987. In the west and south sides of the valley, wells were selected that were screened entirely in either the upper or lower zones. In the east side of the valley, wells were selected that were screened closest to the water table.

Water samples from wells in the west, east, and south sides of the valley were analyzed for triazine herbicides, organophosphate pesticides, and purgeable compounds. Samples from wells in the east side of the valley also were analyzed for chlorophenoxy acid herbicides and carbamate pesticides. Analytical methods and detection limits are described in a report by Wershaw and others (1987).

Triazine herbicides were the only pesticides frequently detected in ground-water samples from the San Joaquin Valley. The areal distribution of wells sampled for triazine herbicides is shown in figure 51. Samples were not analyzed for some compounds, for example, bromacil or diuron. A few samples were analyzed for DBCP and 1,2-dibromoethane, but the detection limit was relatively high (3 µg/L). There-

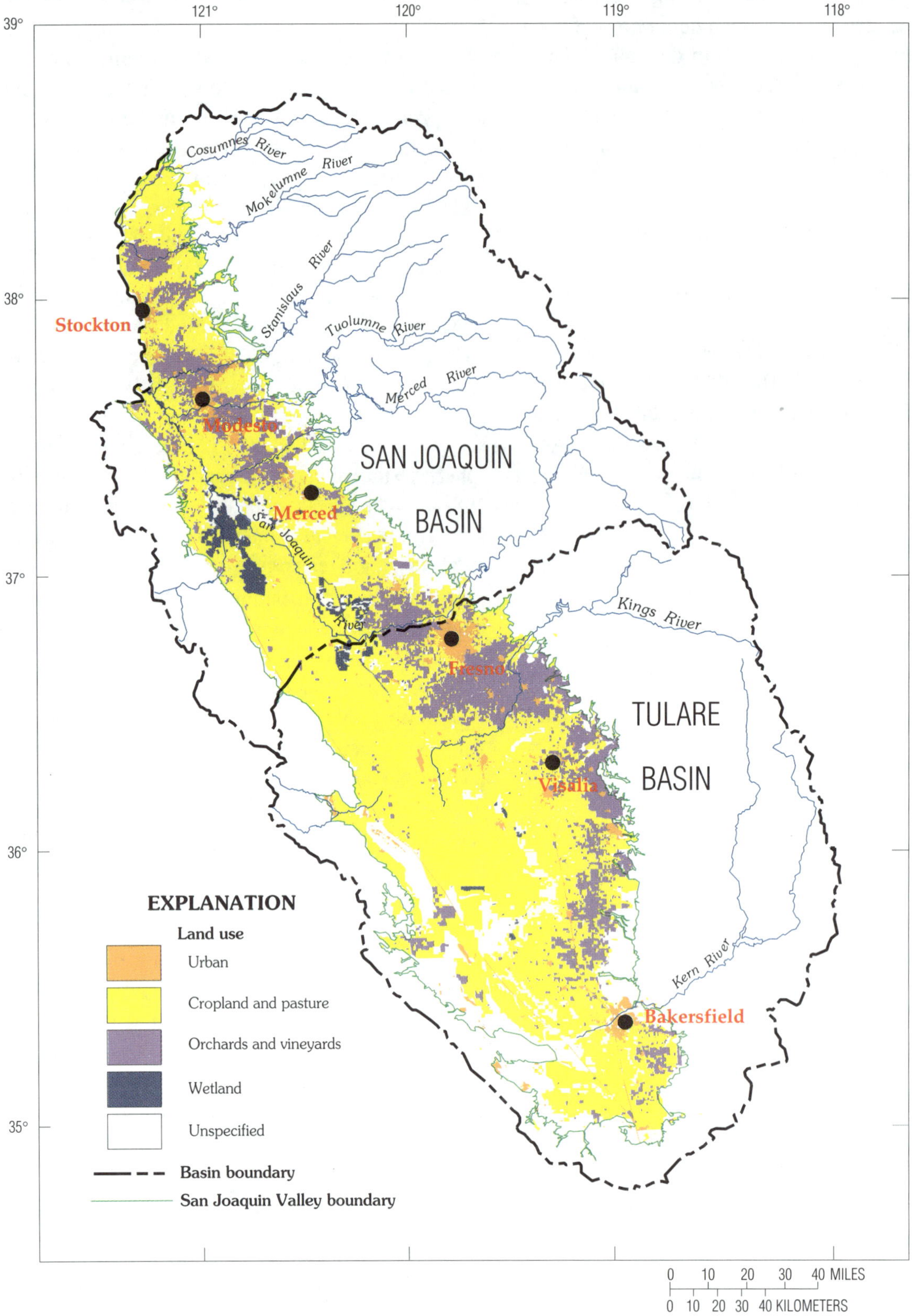

Figure 43. Partial land use in the San Joaquin Valley, California, 1975. (U.S. Geological Survey, Geographical Information Retrieval and Analysis System data base, 1986.)

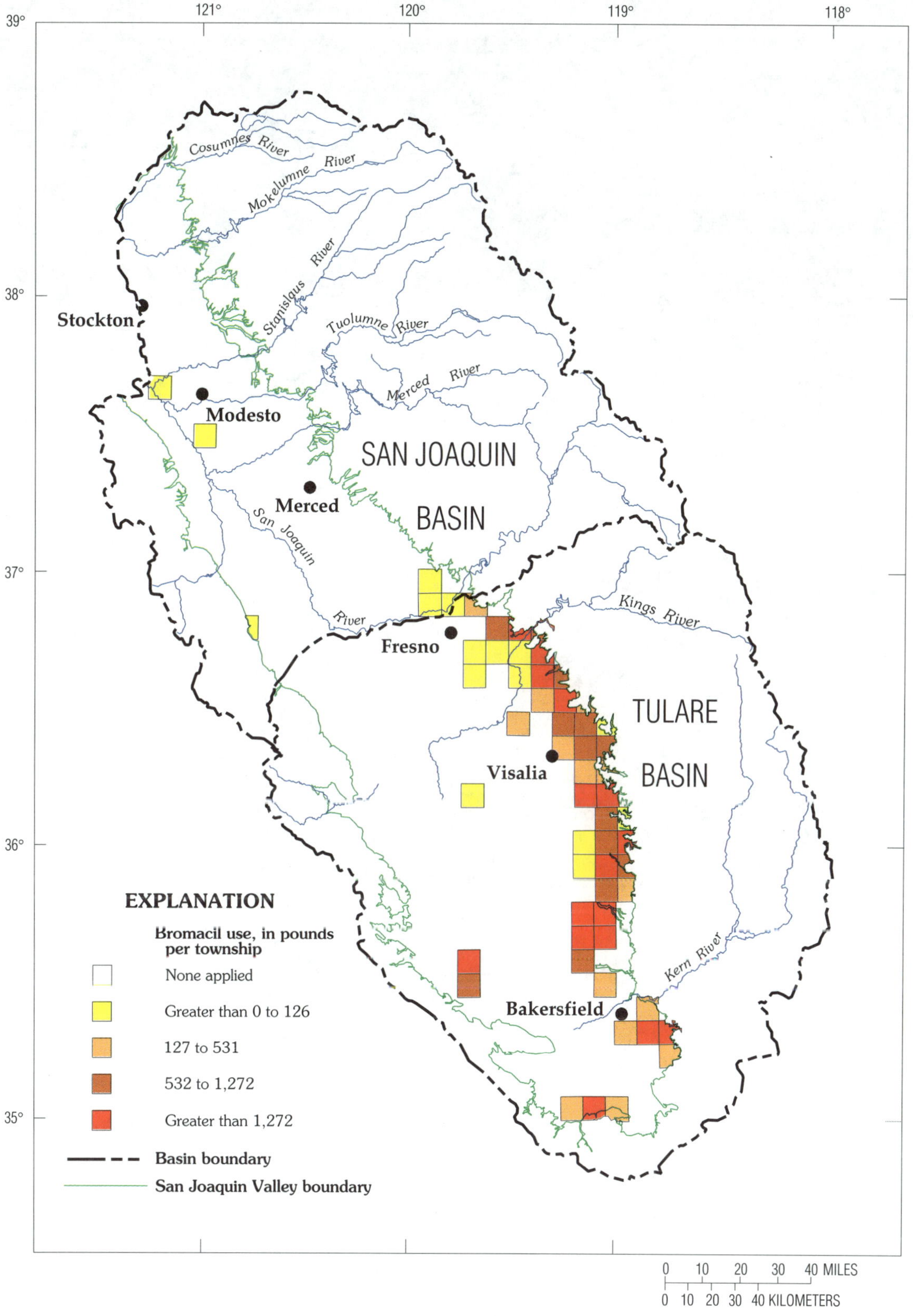

Figure 44. Amount of bromacil applied in the San Joaquin Valley, California, 1990. (Data from California Environmental Protection Agency, Department of Pesticide Regulation well-inventory data base.)

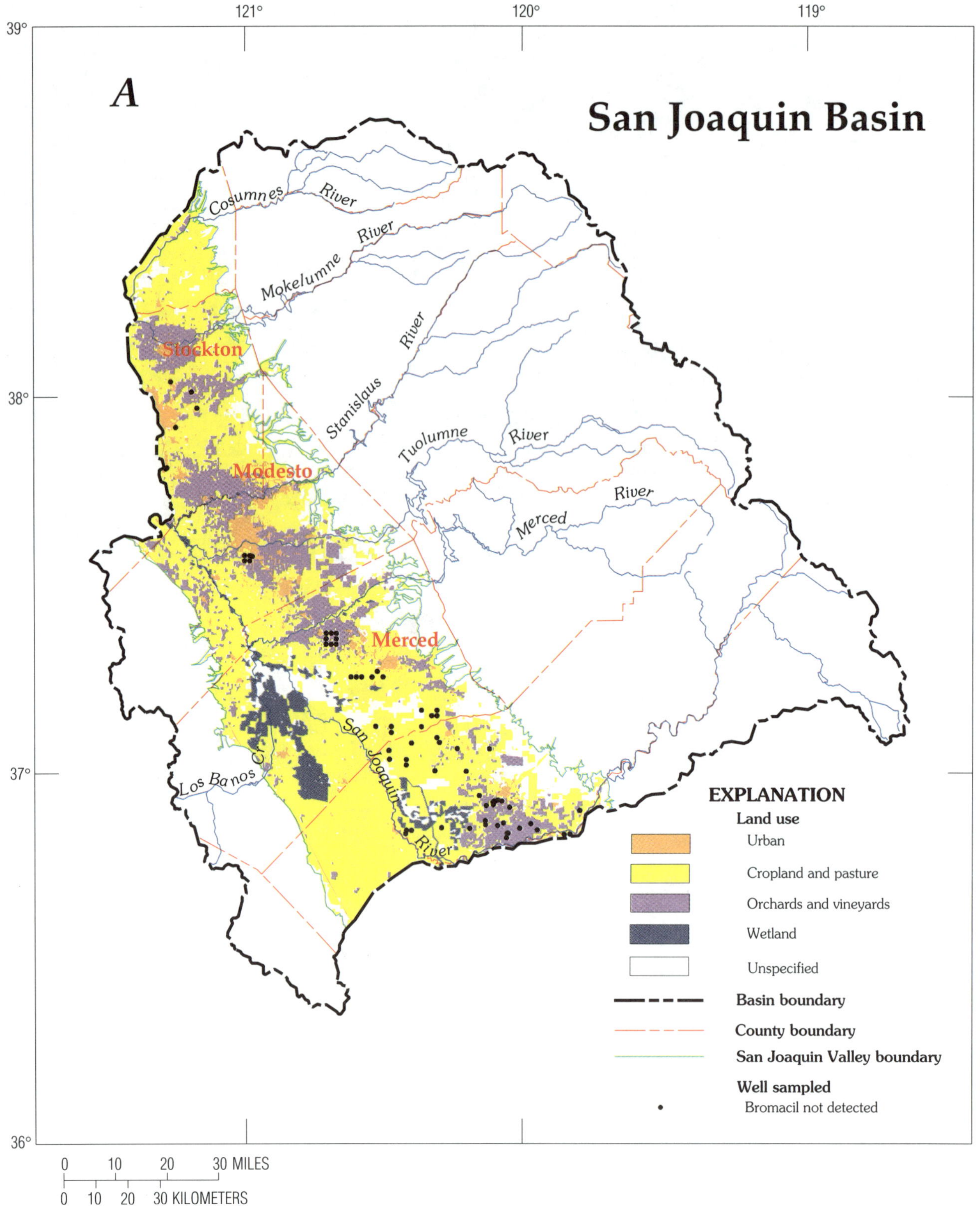

Figure 45. Areal distribution of wells sampled for bromacil, 1971-88, and partial land use, 1975, San Joaquin Valley, California. *A*, San Joaquin Basin; *B*, Tulare Basin. (Data from California Environmental Protection Agency, Department of Pesticide Regulation well-inventory data base, Maes and others, 1991; land use data from U.S. Geological Survey, Geographical Information Retrieval and Analysis System data base, 1986.)

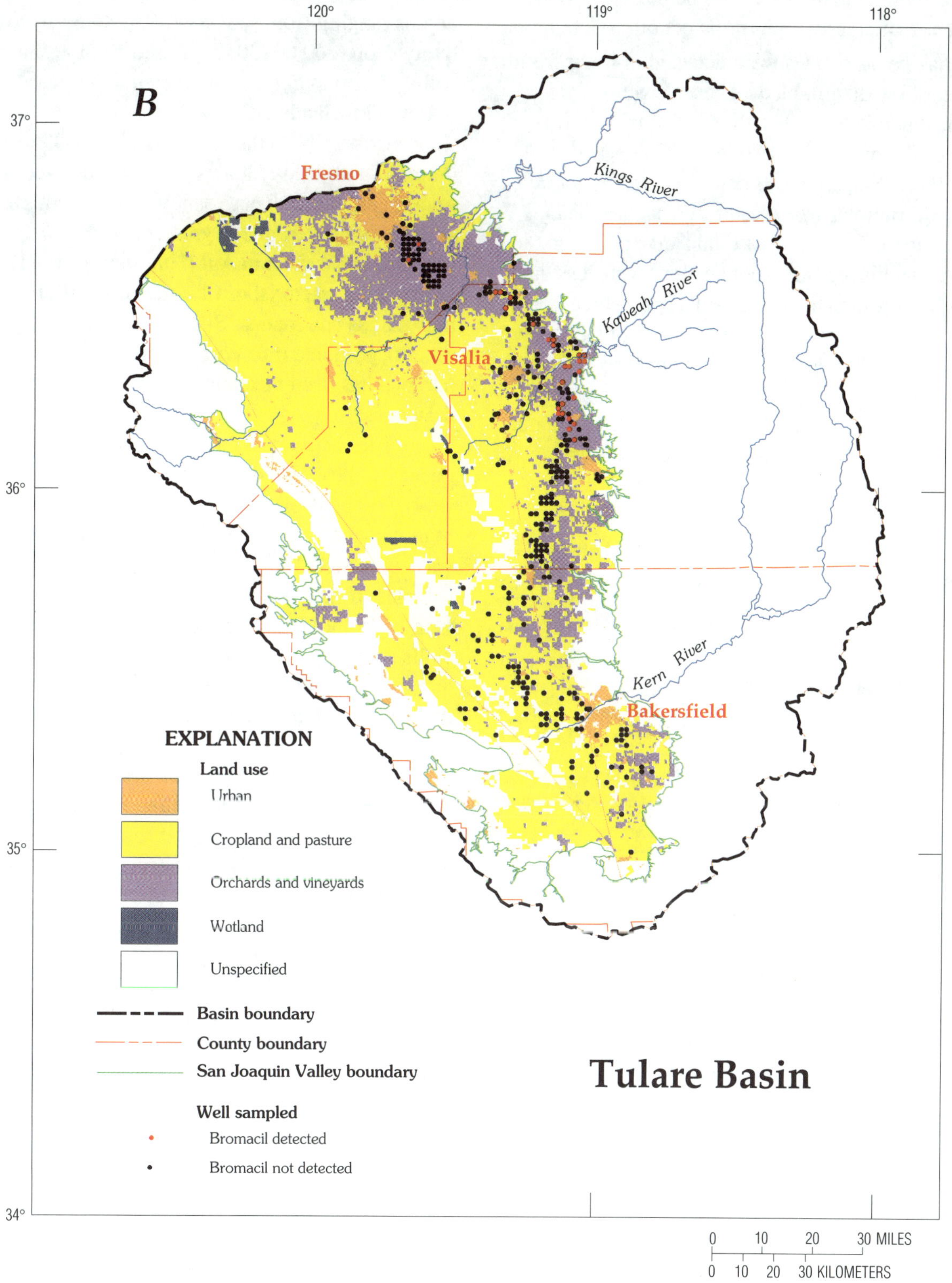

Figure 45—Continued.

fore, it is unlikely that they would be present in concentrations sufficiently high for detection.

Triazine herbicides were detected in water from 25 of the 183 wells sampled. These detections are partly attributable to the half-lives of triazine herbicides in soil, which exceed 1 month (Cohen and others, 1984). Simazine was detected most frequently, primarily in samples from eastern Fresno and Tulare Counties; about 60 percent of total simazine use in the valley was in this region. All detections were less than the drinking water standard for simazine, which is 4 µg/L. Simazine is used on the east side of the valley on vineyards and orchards. Tritium analyses were done on water samples from most wells in the San Joaquin Valley. Although triazine herbicides were not detected in all samples with high tritium concentrations, they were not detected in any samples with no measured tritium. This was expected as the tritium-free water is primarily pre-1952 recharge. Simazine was detected on the east side of the valley in water

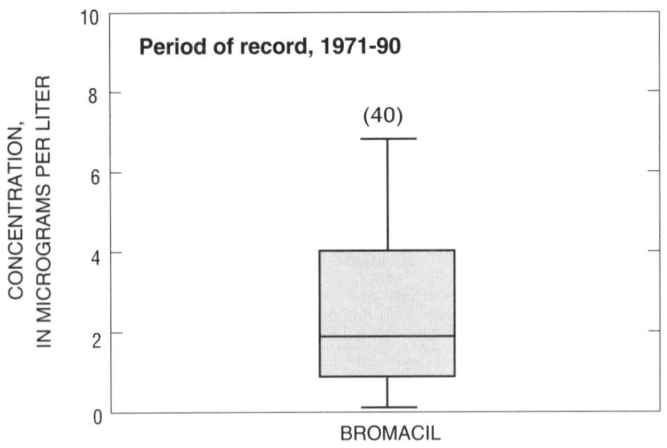

EXPLANATION

Percentile--Percentage of analysis equal to or less than indicated values. (10th and 90th percentiles are not shown for plots with less than 15 data points)

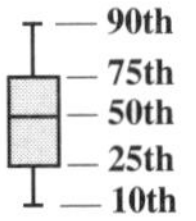

(40) **Number in parentheses indicates number of samples**

Figure 46. Concentrations of bromacil in ground water, San Joaquin Valley, California. (Data from California Environmental Protection Agency, Department of Pesticide Regulation well-inventory data base, Maes and others, 1991. See figure 45 for site locations.)

samples from wells with the midpoint of well-screen depths ranging from shallow (29.85 to 63 ft) to relatively deep (200 to 230 ft). In most of the sampled wells, the midpoint of well-screen depth was about 100 ft below land surface.

Most soils on the east side of the valley are well-drained sands, sandy loams, or loams derived from granitic alluvium. These sandy soils have high permeability (ranging from 2 to greater than 20 in/h) and low total organic carbon in soil (Huntington, 1971). Drainage characteristics of older soils differ from these sands and loams. Silica-cemented hardpan inhibits pesticide transport, but in many areas, the hardpan has been mechanically broken to improve drainage.

The east side of the valley had the highest frequency of pesticide detections of the regional aquifer study. An analysis of land use at each well site of the regional aquifer study on the east side of the valley was completed. Pesticides were detected in water samples from 73 percent of the wells in the orchard/vineyard zones (11 of 15 wells), from 25 percent of the wells in the cropland/pasture zones (16 of 64 wells), and from 15 percent of the wells in the urban zones (2 of 13 wells). Pesticides were not detected in rangeland areas (14 wells).

Simazine concentrations in the soil, sediment, and ground water were measured at a vineyard in eastern Fresno County (Domagalski and Dubrovsky, 1991, 1992). The site was chosen because it is entirely within highly porous dune-sand sediments. Simazine had been detected previously in ground water at this site. These detections were in wells drilled near the water table, which is 42.6 ft below land surface, and as deep as 200 ft below land surface. Measured concentrations of simazine ranged from 0.05 to 0.2 µg/L. Cores of the soil and unsaturated zone were collected 2 months after a normal application of simazine. A concentration of simazine of 2.4 µg/kg was detected in soil near the land surface, but no simazine was detected in other areas of the unsaturated zone. These results indicate that simazine is rapidly transported to the water table following midwinter applications on the vineyard. In contrast, organochlorine compounds, such as endosulfan and DDE, were detected only in the upper soil zone.

Detections of triazine herbicides (atrazine and simazine) in samples from the west side of the valley

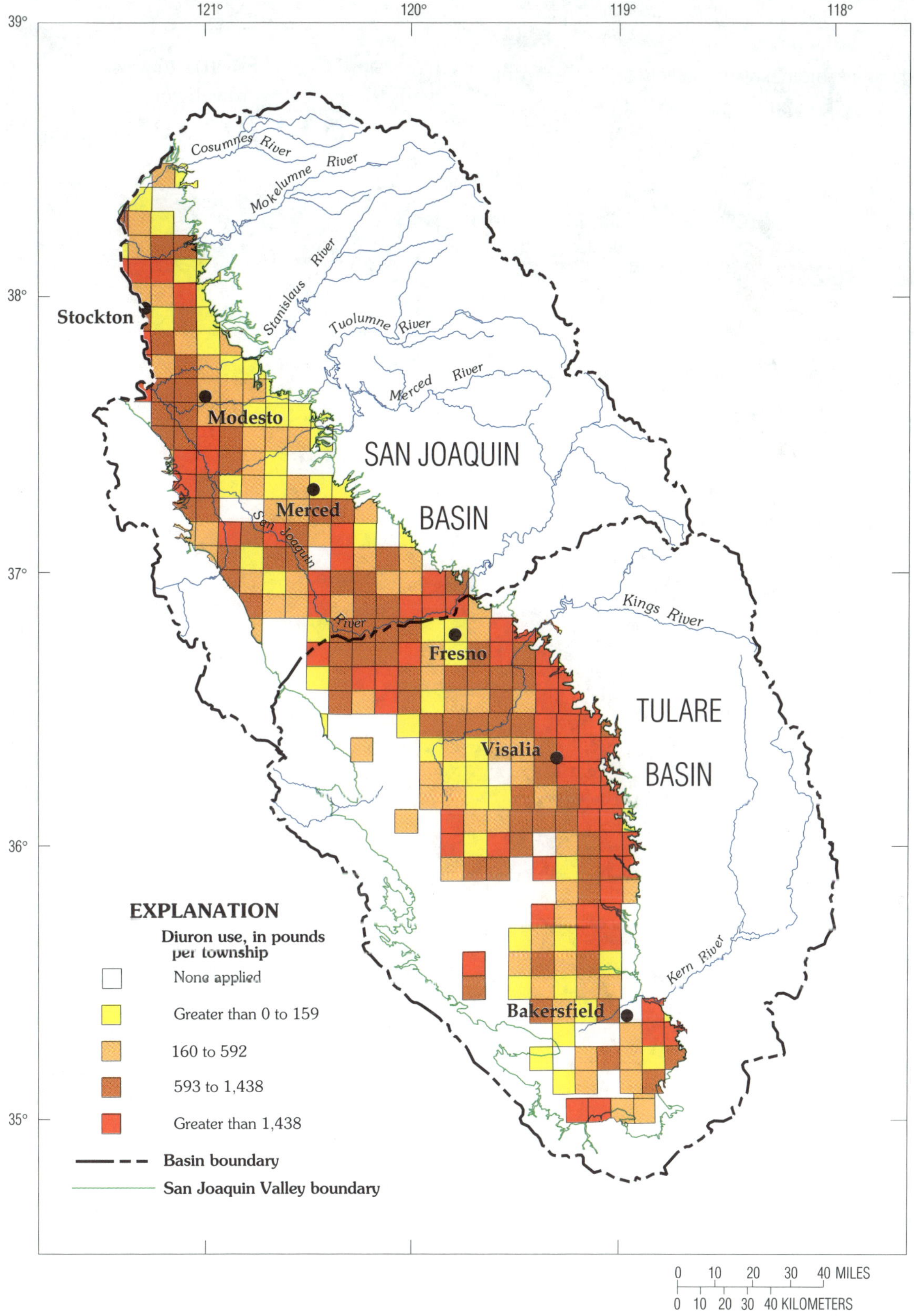

Figure 47. Amount of diuron applied in the San Joaquin Valley, California, 1990. (Data from California Environmental Protection Agency, Department of Pesticide Regulation well-inventory data base, Maes and others, 1991.)

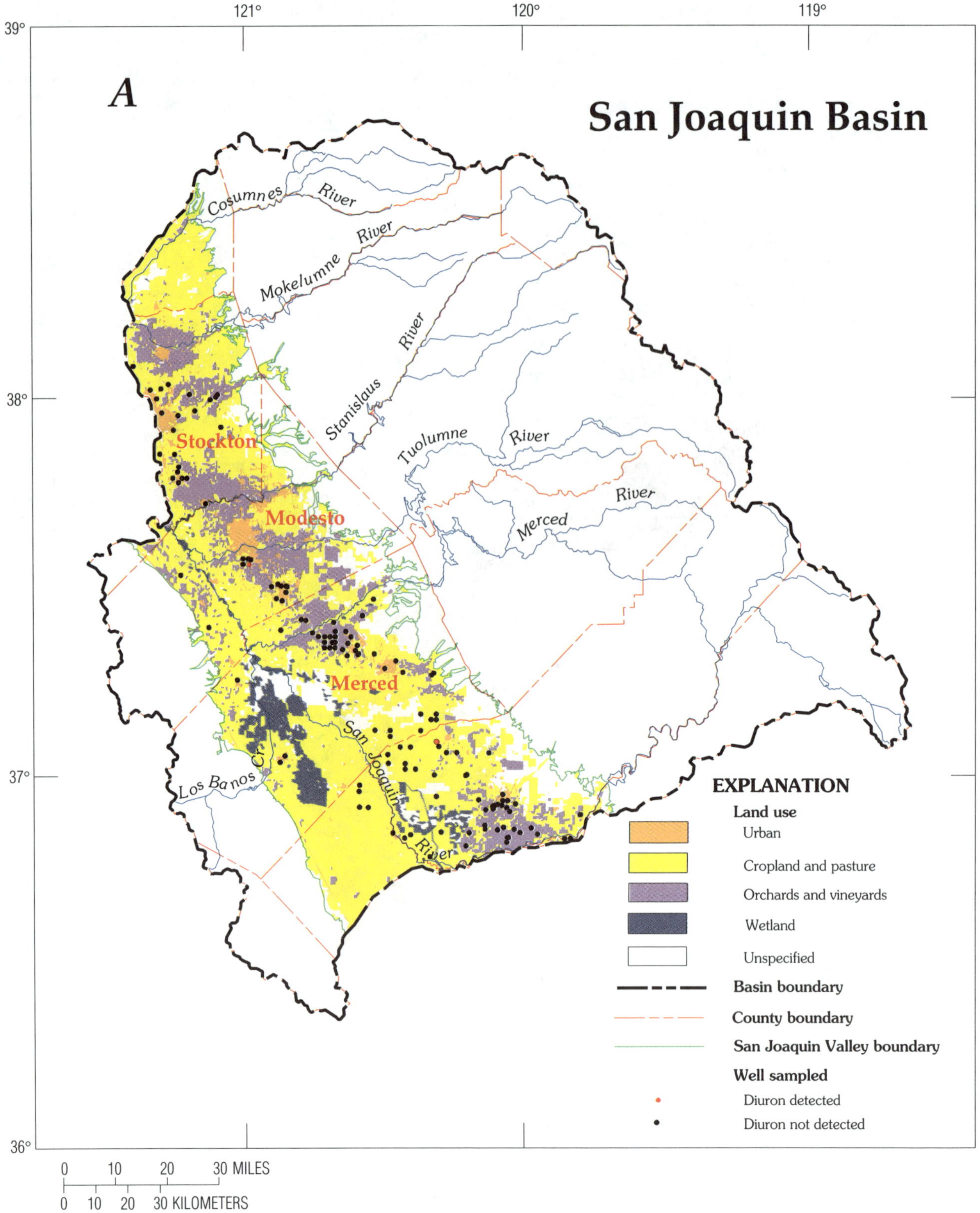

Figure 48. Areal distribution of wells sampled for diuron, 1971-88, and partial land use, 1975, San Joaquin Valley, California. *A.* San Joaquin Basin; *B,* Tulare Basin. (Data from California Environmental Protection Agency, Department of Pesticide Regulation well- inventory data base, Maes and others, 1991; land-use data from U.S. Geological Survey, Geographical Information Retrieval and Analysis System data base, 1986.)

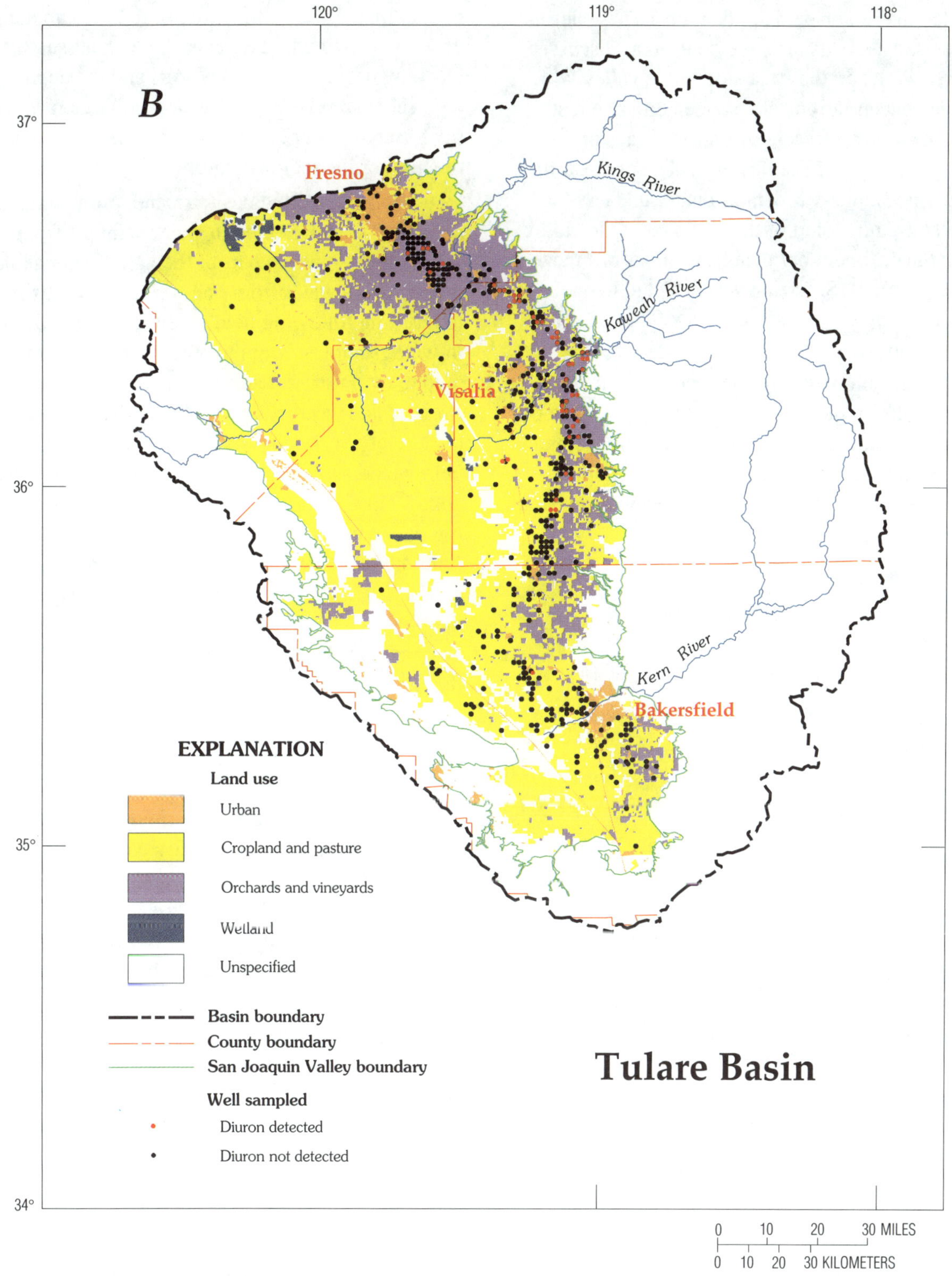

Figure 48—Continued.

are not due to crop or field applications. These herbicides are used as roadside weed control, as discussed previously. Wells on the west side of the valley with detectable concentrations of triazine herbicide residues are in the lower reaches of alluvial fans or the valley trough where the soil is clay. The permeability of the clay is about six orders of magnitude less than the sand and sandy loam of the east side of the valley based on the sediment permeability values of Freeze and Cherry (1979). Simazine was detected in one lower-zone sample (well screened at 207 to 686 ft below land surface).

Atrazine and simazine were detected in water from five wells on the west side of the valley. The tritium content in those samples was high, and the presence of simazine in the lower-zone water sample indicates the possibility of relatively recent water or a contaminated well. It is difficult to attribute the presence of simazine in the lower-zone water sample to leaching because of the clay soils and the depths of the well screens. The ground-water recharge rate probably exceeds the environmental half-life of simazine in this part of the regional aquifer.

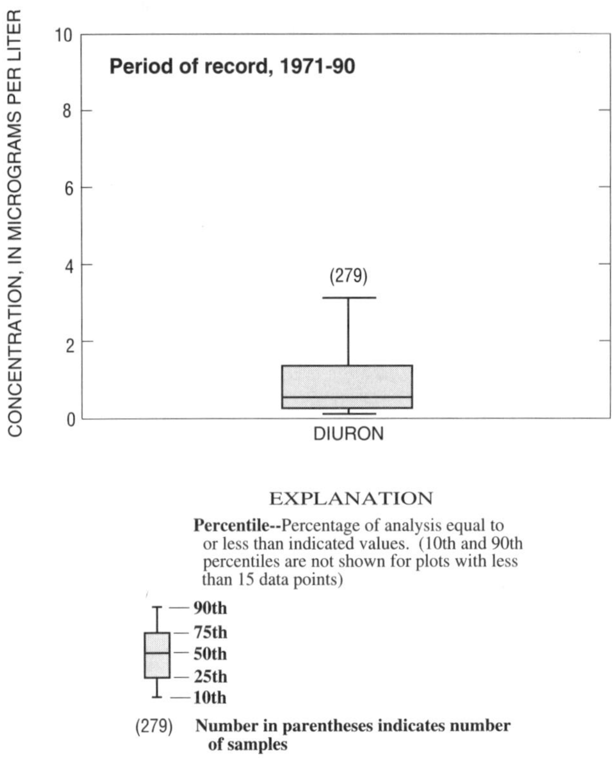

Figure 49. Concentrations of diuron in ground water of the San Joaquin Valley, California. (Data from California Environmental Protection Agency, Department of Pesticide Regulation well-inventory data base, Maes and others, 1991. See figure 48 for site locations.)

Triazine herbicides (atrazine, prometon, and simazine) were detected in water samples from six wells in the southern part of the valley. Simazine detections in water from one well may be attributable to leaching following field applications, but detections in water samples from the other wells cannot be attributed to crop use. Atrazine and prometon are used only for roadside weed control in this area. The six wells have relatively deep screened intervals (192 to 520 ft below land surface); two are screened in the lower zone. Three of the six ground-water samples, including one from the lower zone, were analyzed for tritium; all had detectable tritium, indicating water of recent origin. The detection of triazine herbicides and the presence of tritiated water in these deep wells suggest movement of recent water from near the surface down the well boreholes.

Pesticides in Rain or Foggy Atmospheres

Most rain in the San Joaquin Valley falls during the winter months and coincides with the spraying of dormant almond orchards. As part of a study on urban runoff, rainfall samples were collected at locations in the city of Fresno, California, from December 1981 to March 1983. These rainfall samples were analyzed for the presence of organochlorine and organophosphate pesticides and chlorophenoxy acid herbicides, using the methods described in a report by Wershaw and others (1987). Three organophosphate pesticides were detected—diazinon, malathion, and parathion—and the median concentrations were 0.11, 0.02, and 0.16 μg/L, respectively. Shulters and others (1988) interpreted the presence of these pesticides in rainfall samples to the mode of application in the orchards. The pesticides are applied by high-volume, truck-mounted sprayers, which suspend large quantities of spray in the air, thus facilitating the transport of pesti-

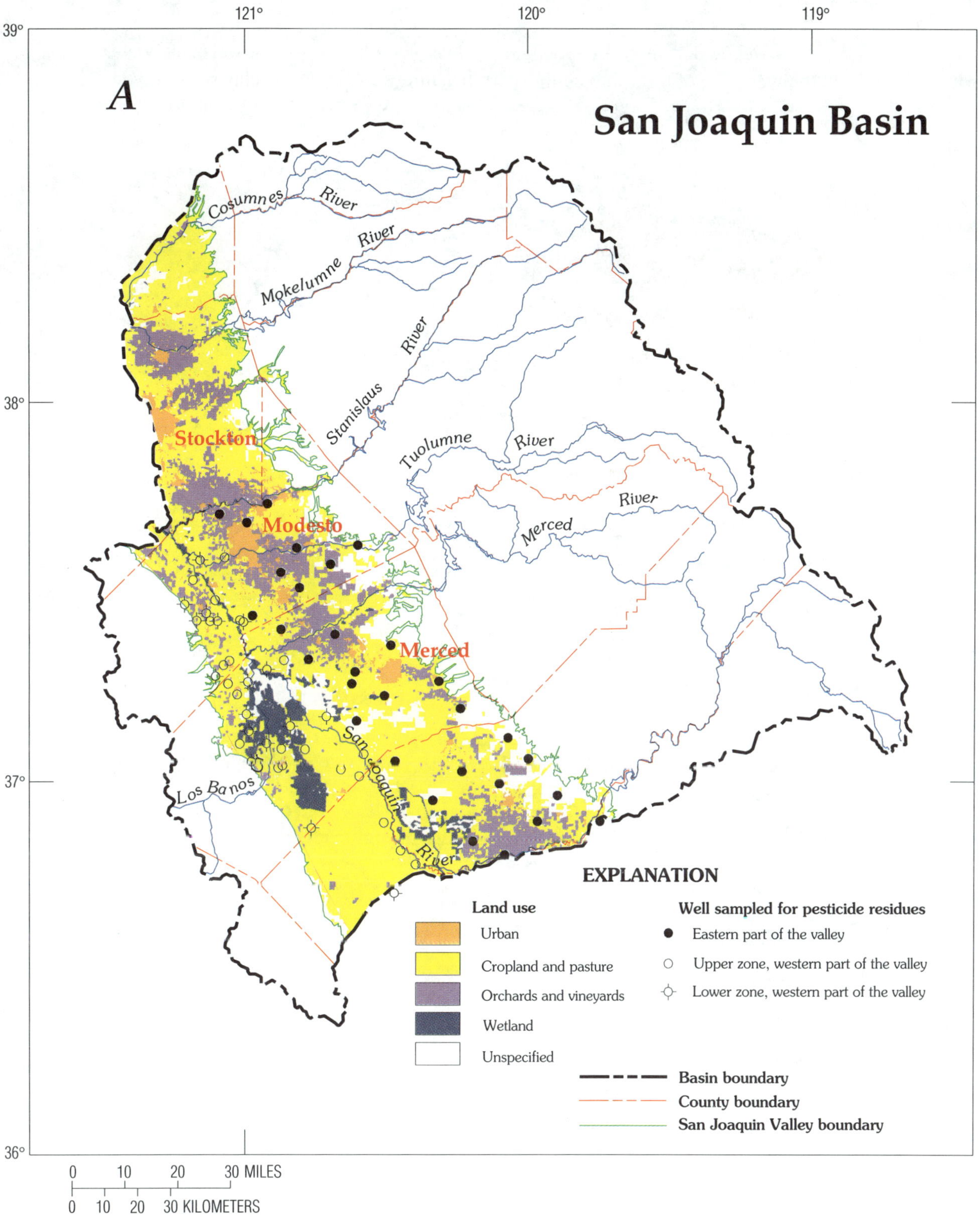

Figure 50. Areal distribution of wells sampled for pesticide residues, San Joaquin Valley, California, 1985-87, and partial land use, 1975. *A,* San Joaquin Basin; *B,* Tulare Basin. (Data from Domagalski and Dubrovsky, 1991, 1992.)

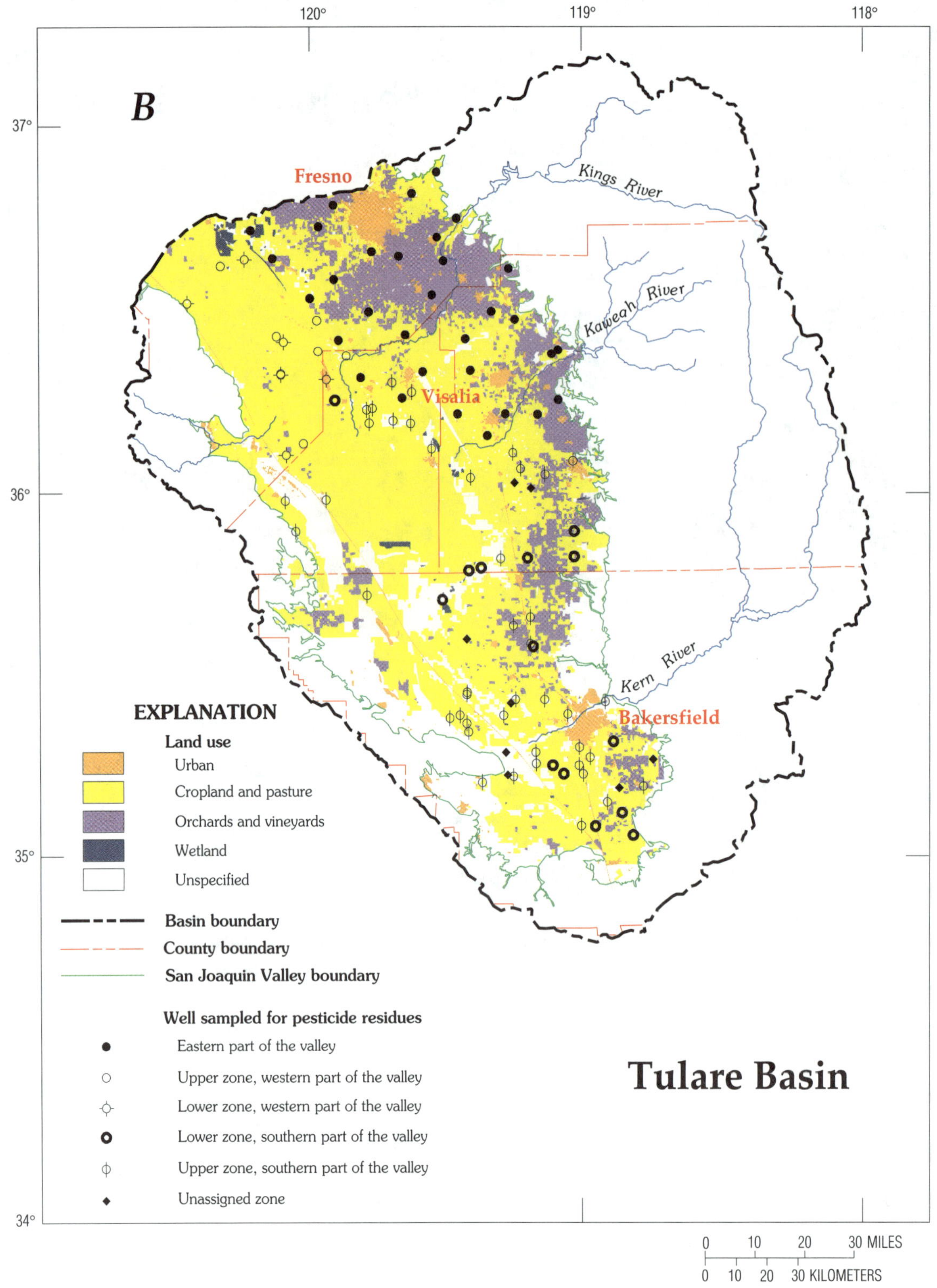

Figure 50—Continued.

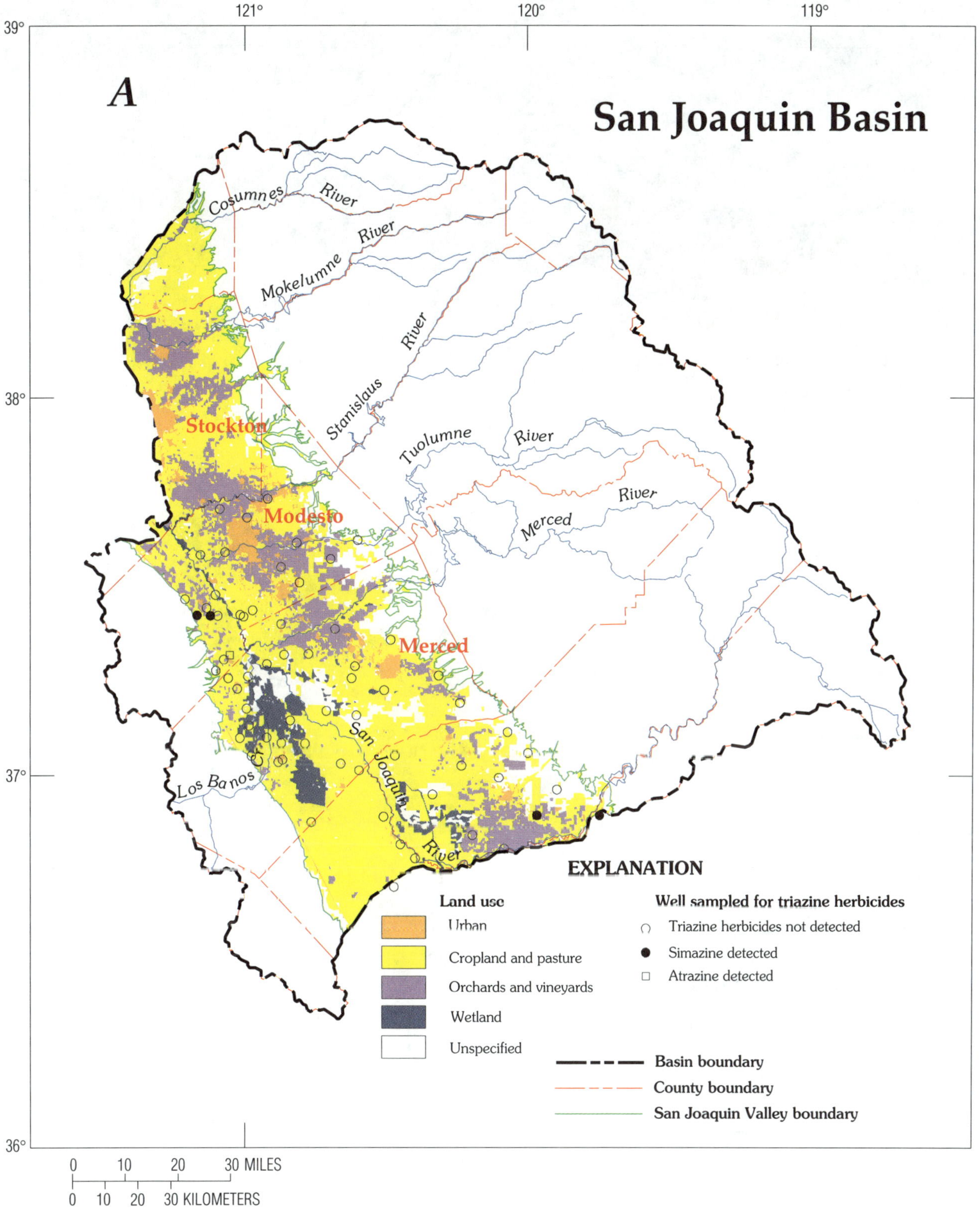

Figure 51. Areal distribution of wells sampled for triazine herbicides, San Joaquin Valley, California, 1985-87, and partial land use, 1975. A, San Joaquin Basin; B, Tulare Basin. (Data from Domagalski and Dubrovsky, 1991, 1992; land-use data from U.S. Geological Survey, Geographical Information Retrieval and Analysis System data base, 1986.)

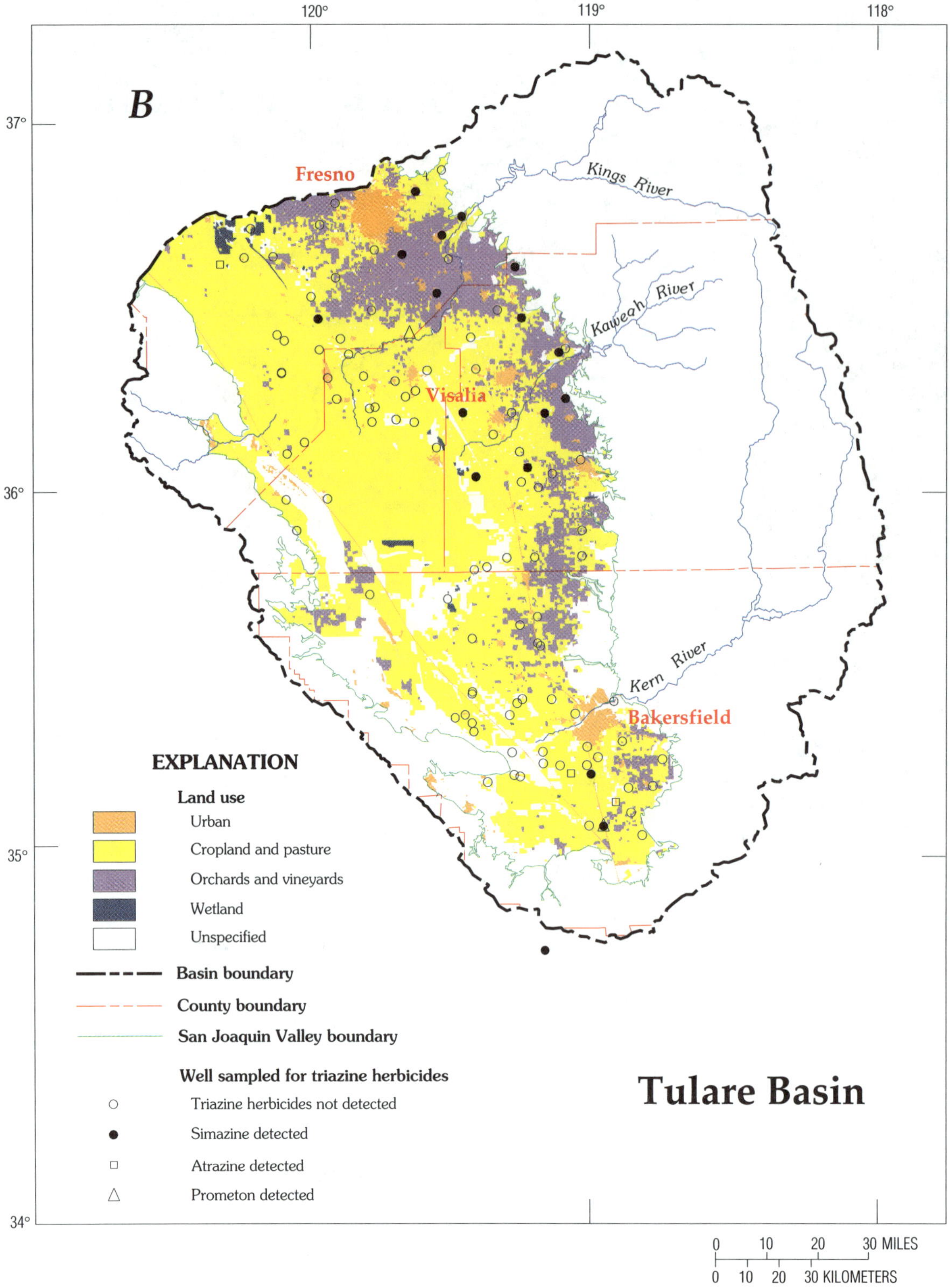

Figure 51—Continued.

cides by wind currents (Shulters and others, 1988). At present, the data are insufficient to assess whether rainfall contributes a significant loading of pesticide residues to surface-water streams.

Fog is common throughout the San Joaquin Valley during November to March; therefore, the fog season overlaps with the dormant spray season. A number of organophosphate pesticides and their oxon derivatives have been detected in fog (Glotfelty and others, 1987, 1990). These pesticides include chlorpyrifos, chlorpyrifos oxon, diazinon, diazinon oxon, methidathion, methidathion oxon, parathion oxon, and parathion. Total concentrations were as high as 60 µg/L for the parent pesticides and as high as 75 µg/L for the oxon derivatives (Glotfelty and others, 1987, 1990). Although these concentrations are high, the importance of fog as a transport mechanism from fields to water bodies has not been assessed.

SUMMARY AND CONCLUSIONS

Although available data sets are limited, 31 pesticides including organochlorine pesticides, organophosphate pesticides, soil fumigants, and triazine herbicides are degrading the quality of surface and ground waters in the San Joaquin–Tulare Basins. Analysis of these preliminary data indicates that, during the dormant-orchard spray season (January to February) and the alfalfa spray season (March to April), some pesticides enter the San Joaquin River and its tributaries. Pesticides also can be detected in the summer, although their sources cannot be assessed at the present time. The data also indicate that the loadings of pesticides are different on tributaries east and west of the San Joaquin River. This is, in part, attributable to soil characteristics. Soils of the east side of the valley are relatively coarser grained than those of the west side. The infiltration rate of water, whether rainfall or irrigation, is greater in the east side of the valley than the west side, and there is less surface-water runoff from fields or orchards. These same soil characteristics make the east side of the valley more susceptible to ground-water contamination than the west side.

Pesticides are applied throughout the year in the study unit, and any monitoring program needs to consider the pesticide application pattern. October is the only month in which pesticide concentrations can be expected to be at their lowest levels. Variations in concentrations, particularly for surface-water sites, are highest during the winter months, the period of highest rainfall in the San Joaquin Valley.

The significance of pesticide detections at surface-water sites is that pesticide concentrations, especially those of the organophosphate pesticide diazinon, are often at toxic levels, as measured by the three-species bioassay of the U.S. Environmental Protection Agency. Toxic contamination can occur in both the winter, when rainfall and runoff occur, and during the irrigation season (May to September). Potentially toxic concentrations were measured most frequently in samples collected from agricultural drains that collect surface runoff. Although pesticides were detected at sites on the larger rivers, such as the San Joaquin River, those concentrations generally were lower than in the agricultural drains. Much of the water-quality data used for this retrospective report is for organochlorine, organophosphate, and carbamate pesticides. Considerably fewer data are available to assess the occurrence and significance of herbicides in surface water.

In general, pesticides were detected more frequently in water samples collected during the irrigation season at sites on the west side of the valley. This greater frequency of occurrence can be attributed to the fine-grained soils of that region. Pesticides enter the drains either dissolved or carried by soil particles. The pesticide detections depend on application patterns and water management. In contrast, pesticides were detected less frequently at surface-water sites east of the San Joaquin River during the irrigation season. These fewer detections probably can be attributed to the coarser grained soils in that area, which lead to more rapid infiltration of irrigation water and less runoff. Pesticides can enter surface water on the east side of the valley during the winter, especially during storms. Pesticides are detected primarily in water samples collected at surface-water sites on the west side of the valley during the time of zero rainfall.

The highest concentrations of organochlorine pesticides, such as DDT and its metabolites, were measured in bed sediments of the western San Joaquin Valley. According to published sources, DDT metabolite concentrations in bed sediments of the San Joaquin River are among the highest of major river systems of the United States.

Triazine herbicides and dibromochloropropane (DBCP) are the pesticides most frequently detected in ground water. DBCP was detected in water samples from more than 1,000 wells, and the median concentration was 2.0 µg/L. This is an order of magnitude greater than the California State drinking-water standard for this compound (0.2 µg/L). Triazine herbicide concentrations were less than the drinking water standards and were not detected as frequently as DBCP. However, monitoring for triazine herbicides has not been as extensive as that for DBCP.

REFERENCES CITED

Amato, J.R., Mount, D.I., Durhan, E.J., Lukasewycz, M.T., Ankley, G.T., and Robert, E.D., 1992, An example of the identification of diazinon as a primary toxicant in an effluent: Environmental Toxicology and Chemistry, v. 11, p. 209-216.

Belitz, Kenneth, and Heimes, F.J., 1990, Character and evolution of the ground-water flow system in the central part of the western San Joaquin Valley, California: U.S. Geological Survey Water-Supply Paper 2348, 28 p.

California Department of Pesticide Regulation, 1988, Pesticide use data: California Department of Pesticide Regulation. [Computer tapes available from California Department of Pesticide Regulation, Sacramento, CA 95814]

———1990, Pesticide use report, indexed by commodity: California Department of Pesticide Regulation, 352 p.

California Division of Mines and Geology, 1959a, Geologic map of California, San Luis Obispo sheet: California Department of Conservation, scale 1:250,000, 2 sheets.

———1959b, Geologic map of California, Santa Cruz sheet: California Department of Conservation, scale 1:250,000, 2 sheets.

———1965a, Geologic map of California, Bakersfield sheet: California Department of Conservation, scale 1:250,000, 2 sheets.

———1965b, Geologic map of California, Fresno sheet: California Department of Conservation, scale 1:250,000, 2 sheets.

California Division of Mines and Geology, 1966, Geologic map of California, San Jose sheet: California Department of Conservation, scale 1:250,000, 2 sheets.

———1967, Geologic map of California, Mariposa sheet: California Department of Conservation, scale 1:250,000, 2 sheets.

———1969, Geologic map of California, Los Angeles sheet: California Department of Conservation, scale 1:250,000, 2 sheets.

Chiou, C.T., Kile, D.E., Brinton, T.I., Malcolm, R.L., Leenheer, J.A., and MacCarthy, Patrick, 1987, A comparison of water solubility enhancements of organic solutes by aquatic humic materials and commercial humic acids: Environmental Science and Technology, v. 21, no. 12, p. 1231-1234.

Clifton, D.G., and Gilliom, R.J., 1989, Sources and concentrations of selenium in the San Joaquin River, *in* Gilliom, R.J., and others, Preliminary assessment of sources, distribution, and mobility of selenium in the San Joaquin Valley, California: U.S. Geological Survey Water-Resources Investigations Report 88-4186, p. 99-113.

Cohen, S.Z., Creeger, S.M., Carsel, R.F., and Enfield, C.G., 1984, Potential for pesticide contamination of groundwater resulting from agricultural uses, *in* Krueger, R.F. and Seiber, J.N., eds., Treatment and disposal of pesticide wastes: American Chemical Society Symposium Series 259, p. 297-325.

Crepeau, K.L., Domagalski, J.L., and Kuivila, K.M., 1994, Methods of analysis and quality-assurance practices of the U.S. Geological Survey Organic Laboratory, Sacramento, California—Determination of pesticides in water by solid-phase extraction and capillary-column gas chromatography/mass spectrometry: U.S Geological Survey Open-File Report 94-362, 17 p.

Crepeau, K.L., Kuivila, K.M., and Domagalski, J.L., 1991, Riverine inputs of pesticides to the Sacramento-San Joaquin Delta, California (abs.): Estuarine Research Federation, 11th Biennial International Estuarine Research Conference, San Francisco, November 10-14, Poster.

Davis, G.H., and Coplen, T.B., 1989, Late Cenozoic paleohydrogeology of the western San Joaquin Valley, California, as related to structural movements in the central Coast Ranges: Geological Society of America, Special Paper 234, 40 p.

Davis, G.H., Green, J.H., Olmsted, F.H., and Brown, D.W., 1959, Ground-water conditions and storage capacity in the San Joaquin Valley, California: U.S. Geological Survey Water-Supply Paper 1469, 287 p.

Davis, G.H., and Poland, J.F., 1957, Ground-water conditions in the Mendota-Huron area, Fresno and Kings Counties, California: U.S. Geological Survey Water-Supply Paper 1360-G, p. 409-588.

Davis, S.N., and Bentley, H.W., 1982, Dating ground water: A short review, *in* Currie, L.A., ed., Nuclear and chemical dating techniques. Interpreting the environmental record: American Chemical Society Symposium Series 176, p. 187-222.

Deeley, G.M., Reinhard, M., and Stearns, S.M., 1991, Transformation and sorption of 1,2-dibromo-3-chloropropane in subsurface samples collected at Fresno, California: Journal of Environmental Quality, v. 20, p. 547-556.

Deverel, S.J., Fio, J.L., and Gilliom, R.J., 1989, Selenium in tile drain water, *in* Gilliom, R.J., and others, Preliminary assessment of sources, distribution, and mobility of selenium in the San Joaquin Valley, California: U.S. Geological Survey Water-Resources Investigations Report 88-4186, p. 77-91.

Deverel, S.J., Gilliom, R.J., Fujii, Roger, Izbicki, J.A., and Fields, J.C., 1984, Areal distribution of selenium and other inorganic constituents in shallow ground water of the San Luis drain service area, San Joaquin Valley, California: A preliminary study: U.S. Geological Survey Water-Resources Investigations Report 84-4319, 67 p.

Domagalski, J.L., and Dubrovsky, N.M., 1991, Regional assessment of nonpoint-source pesticide residues in ground water, San Joaquin Valley, California: U.S. Geological Survey Water-Resources Investigations Report 91-4027, 64 p.

———1992, Pesticide residues in ground water of the San Joaquin Valley, California: Journal of Hydrology, v. 130, p. 299-338.

Foe, C.G., and Connor, Valerie, 1991, San Joaquin watershed bioassay results, 1988-90: California Regional Water Quality Control Board, Central Valley Region, 79 p.

Freeze, R.A., and Cherry, J.A., 1979, Groundwater: Englewood Cliffs, New Jersey, Prentice Hall, 604 p.

Gilliom, R.J., and others, 1989, Preliminary assessment of sources, distribution, and mobility of selenium in the San Joaquin Valley, California: U.S. Geological Survey Water-Resources Investigations Report 88-4186, 129 p.

Gilliom, R.J., and Clifton, D.G., 1990, Organochlorine pesticide residues in bed sediments of the San Joaquin River, California: Water Resources Bulletin, v. 26, no. 1, p. 11-24.

Glotfelty, D.E., Majewski, M.S., and Seiber, J.N., 1990, Distribution of several organophosphorus insecticides and their oxygen analogues in a foggy atmosphere: Environmental Science and Technology, v. 24, p. 353-357.

Glotfelty, D.E., Seiber, J.N., and Liljedahl, L.A., 1987, Pesticides in fog: Nature, v. 325, p. 602-605.

Greeson, P.E., Ehlke, T.A., Irwin, G.A., Lium, B.W., and Slack, K.V., eds., 1977, Methods for collection and analysis of aquatic biological and microbiological samples: U.S. Geological Survey Techniques of Water-Resources Investigations, book 5, chap. A4, 332 p.

Guy, H.P., and Norman, V.W., 1970, Field methods for measurement of fluvial sediment: U.S. Geological Survey Techniques of Water-Resources Investigations, book 3, chap. C2, 59 p.

Hirsch, R.M., Alley, W.M., and Wilber, W.G., 1988, Concepts for a National Water-Quality Assessment Program: U.S. Geological Survey Circular 1021, 42 p.

Huntington, G.L., 1971, Soil survey, eastern Fresno area, California: Washington, D.C., U.S. Department of Agriculture, Soil Conservation Service, 323 p.

Johnson-Logan, L.R., Broshears, R.E., and Klaine, S.J., 1992, Partitioning behavior and mobility of chlordane in groundwater: Environmental Science and Technology, v. 26, p. 2234-2239.

Laudon, Julie, and Belitz, Kenneth, 1991, Texture and depositional history of late Pleistocene-Holocene alluvium in the central part of the western San Joaquin Valley, California: Bulletin of the Association of Engineering Geologists, v. 28, no. 1, p. 73-88.

Maes, C.M., Pepple, Mark, Troiano, J., Weaver, D., and Kimaru, W., 1991, Sampling for pesticide residues in California well water, 1991 update, well inventory data base: California Environmental Protection Agency, Department of Pesticide Regulation, 185 p.

Mount, D.I., 1989, Methods for aquatic toxicity identification evaluations: Phase III. Toxicity confirmation procedures: Duluth, Minnesota, U.S. Environmental Protection Agency Environmental Research Laboratory, EPA 600/3-88-036.

Mount, D.I. and Anderson-Carnahan, Linda, 1988, Methods for aquatic toxicity identification evaluations: Phase I. Toxicity characterization procedures: Duluth, Minnesota, U.S. Environmental Protection Agency Environmental Research Laboratory, EPA 600/3-88-034.

———1989, Methods for aquatic toxicity identification evaluations: Phase II. Toxicity identification procedures: Duluth, Minnesota, U.S. Environmental Protection Agency Environmental Research Laboratory, EPA 600/3-88-035.

Nady, Paul, and Larragueta, L.L., 1983, Estimated average annual streamflow into the Central Valley of California: U.S. Geological Survey Hydrologic Investigations Atlas, HA-657, scale 1:500,000, 1 sheet.

Page, R.W., 1986, Geology of the fresh ground-water basin of the Central Valley, California, with texture maps and sections: U.S. Geological Survey Professional Paper 1401-C, 54 p.

San Joaquin Valley Drainage Program, 1990, A management plan for agricultural subsurface drainage and related problems on the westside San Joaquin Valley: Final Report: U.S. Department of the Interior, 183 p.

Shulters, M.V., Oltmann, R.N., and Grabbe, R.R., 1988, Pesticides in rainfall samples collected at Fresno, California, December 1981 through March 1983, *in* Subitzky, Seymour, ed., Selected Papers in the Hydrologic Sciences, 1987: U.S. Geological Survey Water-Supply Paper 2330, p. 35-40.

Troiano, J.J., and Segawa, R.T., 1987, Survey for herbicides in well water in Tulare County: California Department of Food and Agriculture, Environmental Hazards Assessment Program, 13 p.

U.S. Bureau of Census, 1990, 1990 census of population and housing: U.S. Department of Commerce, Public Law 94-171, on CD-ROM.

U.S. Geological Survey, 1978, Hydrologic unit map—1978, State of California: U.S. Geological Survey Hydrologic Unit Map, scale 1:500,000, 2 sheets.

———1986, Land use and land cover digital data from 1:250,000 and 1:1,000,000-scale maps: U.S. Department of Interior, National Mapping Program, Technical Instructions, Data Users Guide 4, 36 p.

Wershaw, R.L., Fishman, M.J., Grabbe, R.R., and Lowe, L.E., 1987, Methods for the determination of organic substances in water and fluvial sediments: U.S. Geological Survey Techniques of Water-Resources Investigations, book 5, chap. A3, 80 p.

Williamson, A.K., Prudic, D.E., and Swain, L.A., 1989, Ground-water flow in the Central Valley, California: U.S. Geological Survey Professional Paper 1401-D, 127 p.

☆U.S. GOVERNMENT PRINTING OFFICE: 1997–646-484